Technology and Society

Technology and Society: A World History explores the creative power of humanity from the age of stone tools to the digital revolution. It introduces technology as a series of systems that allowed us to solve real-world problems and create a global civilization. The history of technology is also the history of the intellectual and cultural place of our tools and devices. With a broad view of technology, we can see that some of the most powerful technologies such as education and government produce no physical object but have allowed us to coordinate our inventive skills and pass knowledge through the ages. Yet although all human communities depend on technology, there are unexpected consequences from its use which, as Ede shows, form a crucial part of this rich story.

Andrew Ede is a historian of science and technology whose research focuses on the intersection of science and technology in the twentieth century. He is the Director and founder of the Science, Technology and Society Program at the University of Alberta. Professor Ede earned his doctorate at the Institute for the History and Philosophy of Science and Technology at the University of Toronto. With Lesley B. Cormack, he is the co-author of the best-selling *A History of Science in Society*. He is a member of the Canadian Society for the History and Philosophy of Science, the History of Science Society, and the Society for the History of Technology, and was the program coordinator for the Three Societies Meeting in 2016.

A World History

Andrew Ede
University of Alberta

CAMBRIDGE
UNIVERSITY PRESS

CAMBRIDGE
UNIVERSITY PRESS

Shaftesbury Road, Cambridge CB2 8EA, United Kingdom

One Liberty Plaza, 20th Floor, New York, NY 10006, USA

477 Williamstown Road, Port Melbourne, VIC 3207, Australia

314–321, 3rd Floor, Plot 3, Splendor Forum, Jasola District Centre, New Delhi – 110025, India

103 Penang Road, #05–06/07, Visioncrest Commercial, Singapore 238467

Cambridge University Press is part of Cambridge University Press & Assessment, a department of the University of Cambridge.

We share the University's mission to contribute to society through the pursuit of education, learning and research at the highest international levels of excellence.

www.cambridge.org
Information on this title: www.cambridge.org/9781108441087

DOI: 10.1017/9781108348539

First published 2019

A catalogue record for this publication is available from the British Library

Library of Congress Cataloging-in-Publication data
NAMES: Ede, Andrew, author.
TITLE: Technology and society : a world history / Andrew Ede.
DESCRIPTION: Cambridge : Cambridge University Press, 2019. |
Includes bibliographical refernces and index.
IDENTIFIERS: LCCN 2018033892| ISBN 9781108425605 (hardback) |
ISBN 9781108441087 (paperback)
SUBJECTS: LCSH: Technology – Social aspects.
CLASSIFICATION: LCC T14.5 .E37 2019 | DDC 303.48/3–dc23
LC record available at https://lccn.loc.gov/2018033892

ISBN   978-1-108-42560-5   Hardback
ISBN   978-1-108-44108-7   Paperback

Contents

Maps

Figures

Tables

Preface

The history of technology represents one of the most important thematic approaches to world history. It takes us from our ancient ancestors to the modern day, from stone tools and the discovery of fire to global transportation systems and supercomputers. Regardless of the vast increase in the complexity of the tools we have available to us today, our relationship with technology remains the same: We use technology to solve real-world problems.

Technology represents some of the greatest achievements of the human mind and made some of the darkest moments in history possible. Despite its importance to the study of history, there have been very few texts available for instructors and students at the undergraduate level. After many years of teaching the history of technology, in this book I provide a synthesis of various ideas and approaches to the question of technology as a key component of world history, based on the argument that technology is a system, not a collection of artifacts. Every form is imbedded in human society and requires human action to come into being, find use and in some cases be discarded. A hammer is only a hammer because it was created to do a particular thing, but both the maker and the user must have an understanding of what a hammer is for to make it useful. The hammer, by itself, is not the technology. It is the melding of the utility of the artifact and the conception of it that make a hammer an object of technology. There are also forms of technology that are not based on physical objects. Education and government are examples of these "invisible technologies." Education is one of the most powerful technologies ever created, in part because it trains people to use technology.

The text identifies and discusses some of the pivotal moments in world history that have a technological component. It introduces a number of philosophical ideas that are important to thinking about technology, such as technological determinism and the problem of resistance to technological change.

I would like to thank my partner in everything Lesley Cormack for her help with this project. I would also like to thank my students, particularly those in the Science, Technology and Society Program, who inspired the creation of this book. In addition, the book benefited from the kind assistance of the reviewers whose encouragement and helpful suggestions improved the text.

1 Introduction: Thinking about Technology

Humans have always used technology, so a history of humanity would be incomplete without understanding the role it has played in our collective story. To understand technology in society, we need a definition of technology. This is not a simple task, since thinkers have been debating its effect on society since the time of the ancient Greeks. Technology should not be confused with tools, devices or machines. Physical objects, whether a stone axe or a supercomputer, are created in a specific human context and are part of a system of human knowledge. With a working definition, we can look at several of the most important issues that scholars have raised about the interaction of technology and society, particularly the problem of technological determinism, the general conditions that contribute to invention and the potential problems of technology in society.

To be human is to use technology. Everything we do, from telling stories around a campfire to examining the farthest reaches of the universe, is done using technology. The web of technology that makes human life possible is so pervasive that we are often only aware of it when it breaks down or suddenly changes. It is so closely tied to human existence that we identify groups of people by their access to technology, comparing "industrialized" countries to "developing" countries. We even classify vast periods of human history on the basis of technology such as the Neolithic period or "New Stone Age," followed by the "Bronze Age."

Some scholars have described technology as the ability to make tools, while others see it as a kind of framework that surrounds us. This book argues that technology does not exist on its own as something separate from people and the societies we create. At a fundamental level, we are our technology. In other words, a history of technology is a history of the development of human society and, as such, this book places in the foreground the social context of technology. This does not mean that tools are unimportant. No history of technology can ignore the history of invention, but inventions, no matter how marvelous, are always created within the context of the society of the inventor. Historically, the success or failure of a particular tool or device was determined not simply by the quality or utility of the invention itself, but by a range of social factors such as the degree to which society is open to change or the social status of the inventor. Since no history of technology could encompass the vast range of inventions and their use, I have selected examples that illustrate this integral relationship of technology and society and that were key to transforming human history.

⚙ Technology: A Definition

The main difficulty with defining the term "technology" is that its common use is vague and implies value. In everyday use the term means the created physical objects around us, and there is a strong sense that new technology is better than existing technology. We often present objects as the technology, whether it is in an advertisement for the latest tablet computer, or the newest "technology of hair care" shampoo. There also seems to be a special category of "high technology." In advertising, "high tech" is always the most advanced – and the implication is that it is the best technology. This suggests there is a hierarchy, with low technology (usually old or requiring manual operation) at the bottom and high technology (usually electronic and increasingly autonomous) at the top.

The problem with using common meanings for complex terms is that they can cloud our understanding of what is really happening. Technology is not the object itself. An airplane or a spoon is only an artifact or a product of human construction and craft, and is thus the physical component of technology, but it is not the technology itself. It is only when the artifact is used that it becomes something more than a collection of matter.

A more precise definition of technology is that it is the *system* by which we attempt to solve real-world problems. In other words, technology presents the complex web of knowledge, social connections and behavior that makes it possible for us to solve real-world problems. Most of the time technology includes a material object that we use to interact with the environment, but not all technologies require a physical artifact. To understand why technology must be a system, consider an incandescent light bulb. The light bulb is a simple object with no moving parts and consisting of glass, a bit of metal and some ceramics. As a physical object, it could have been built by the ancient Egyptians, but it was not invented until the late nineteenth century when the controlled production of electricity became possible. The light bulb in your home cannot be used for its designed purpose without access to an electrical network that includes the generating plant, the delivery system and the people who create and maintain the network. Although a person with moderate knowledge of glassmaking, smelting and ceramics could make a single working light bulb, the incandescent bulb is really a manufactured object, and as such is part of a vast manufacturing chain that links mines, refineries, transportation, factories and retail operations together. The light bulb, humble in creation and simple in construction, is in fact part of one of the most extensive and complex systems every devised. To use a light bulb is to take part in the vast system needed to make the light bulb possible. Without the access to the system, the bulb is an artifact or a manufactured object, not the technology itself.

The distinction between technology and artifact is important because it is possible to have a technology that does not require physical objects or produces no artifacts. For example, education is one of the greatest technologies we have ever created, but it does not produce a material end product the way that we produce automobiles or sewing machines. These "invisible technologies" include language, education and forms of governance and extend to such things as national governments, corporations and sports governing bodies.

Some artifacts seem to be self-contained and their purpose self-evident. A knife, whether it is a stone tool from the Paleolithic era or a surgical steel blade manufactured today, seems like it should be understood as a cutting tool. Yet even the simplest objects had to be invented and their use taught to new generations. A modern surgical steel scalpel would be instantly recognized by a Babylonian surgeon accustomed to bronze cutting tools, but a heart regulating pacemaker would be a mystery and completely unusable, even if it could be transported back in time. In a thought experiment, we could bring a Babylonian surgeon to the present and train him to use a pacemaker, but this would mean that our ancient surgeon would become part of the web of knowledge in which the pacemaker was embedded. The pacemaker fits within the context of our current society with all its education, infrastructure and intellectual concepts. Our current medical technology includes the knife and the pacemaker, but the Babylonian surgeon's world included the knife, but not anything electrical.

It is important to remember that the reverse is also true. To really understand the world of the Babylonian surgeon, we must learn about the network and social context that made his technology possible. Thus, the Babylonian knife and the modern scalpel both function as cutting devices because our society shares with the Babylonians the concept of surgery, but to understand what the knife means and how it was used by the ancient surgeon, we need to understand the education and social circumstances of the surgeon, not just note the existence of a type of knife. Both societies had real-world problems and created a device that would solve those problems. Assuming that our knowledge is automatically superior to or subsumes the knowledge of people of the past can lead us to undervalue our ancestors and misinterpret history.

❦ The Concept of Invention

Whether a knife is made of bronze or high carbon steel, a knife is only a knife if the user understands the object as part of a larger category of things that can cut. The concept of cutting must already exist in the mind of the person taking the action. To create such a plan requires the ability to identify a problem, conceive of a desired end point and take action to reach the desired conclusion. When the path from problem to solution does not already exist, we call the act of creating a new path "invention." Like technology, most inventions involve the creation of artifacts, but not all inventions are physical.

The story of invention has been one of the most enduring and popular forms of history. We have created entire museums dedicated to important inventions such as automobiles, aircraft, weapons of war, ceramics and shoes. Our interest has spanned the ages. In ancient China, Sima Qian (c. 145–90 BCE) wrote about many of the great inventions in his *Historical Record*. Hero of Alexandria (c. 10–70 CE) wrote about his inventions, including a type of steam engine called the *aeolipile*. His work was rediscovered by Islamic scholars around 1000 CE and then by Europeans during the Renaissance and in both cases contributed to periods of new creativity. The *Encyclopédie* (1765–72) published by Denis Diderot (1713–84), was one of the greatest documents of the Age of Enlightenment, and contained some of the most detailed descriptions of tools and machines ever published. Today almost every important invention has its own book. Even something as humble as the screw was examined by Witold Rybczynski in his book *One Good Turn: A Natural History of the Screwdriver and the Screw* (2000).

Part of the reason that inventions get so much attention is that there is a link between the tools a society has and the power of that society. Thus, our interest in the history of invention extends far beyond a fascination with the devices themselves and becomes a study of the course of human history, particularly the dramatic moments such as the clash of nations, the rise and fall of industries and other dramatic turning points.

Although devices can be looked at without direct reference to the creators of those devices, historians have also been fascinated by inventors. The stories of inventors are often a significant part of history, particularly when some new thing (a cannon, a radio, a supposedly unsinkable ship) can be pointed to as "changing the course of history." People like Archimedes, Li Jun, Johann Gutenberg or Thomas Edison become icons, often woven into national identity and serving as role models. This is particularly true of inventors who faced opposition or deprivation in the course of creating world-changing devices. Stories such as Archimedes running naked through the market shouting "Eureka!," Johann Gutenberg forced to sell his press and dying in poverty, John Harrison (inventor of the first reliable marine chronometer) denied the prize money for his invention for many years, have fascinated people for generations. These stories are often presented as morality lessons teaching us things like the value of curiosity and persistence, that individuals can make a difference, especially if they are resolute, or that geniuses are often unappreciated in their own time. In the case of many of the greatest inventions, the power to change the course of history was unexpected, adding to the romance of the story.

Although this book will look at many inventors and their inventions, there is a difference between a history of human invention and a history of technology. A history of invention treats tools and devices as separate from the world in which they are created and often attempts to trace a kind of family tree of development. It portrays inventors as heroic figures (who may also be misunderstood or tragic) who exist outside the norms of society. This is not to say that there weren't great inventors, but rather that the act of invention is not what creates a new technology. Successful invention requires the adoption of the new technology and that requires a collaboration of intellectual insight, technical utility and social acceptance. It is important to remember that there are many examples of inventions that were not adopted even though they were at some technical level better than their competition, and conversely, inventions that were far less useful than what already existed and yet were adopted. There were also periods when societies resisted adopting any major new invention, and others that have become known as periods of great inventiveness.

Progressivism and Presentism

The belief that there is a kind of arrow of development from primitive to highly developed is called "progressivism." A related belief called "presentism" is the idea that the past existed only to produce us in the present, and that the past can be judged by the standards and knowledge of the present. Historians try to avoid these two "isms," but it can be a particularly difficult job for historians of technology. In consumer culture, especially as it has developed in the industrialized world, there is constant pressure to produce new and "improved" models, partly because it is the job of engineers and designers to create new things, and partly to entice consumers to continue to purchase products. Thus, any historical examination of consumer goods strongly suggests that there *is* a hierarchy of product quality from rudimentary devices produced in the past to the improved versions available today, and toward the glorious and almost magical products on the drawing boards of scientists and engineers for the future. The transition from crystal radios to radios with vacuum tube technology to black and white television, to color television, to digital high-definition LED flat screen televisions with stereo surround sound seems like a perfect example of the improvement of technology. Although people frequently complain about the decline of the quality of consumer goods and the disappearance of craft skill, for the most part the material goods of the present are superior to those produced in the past in terms of reliability, price and availability.

Although we can argue for a kind of progressivism in technology, it would be wrong to see human history being based on an ideology of progressivism. The conception of constant material progress as a component of society is an ideology found primarily in modern Western culture (meaning Europe and European-settled regions including North America, South America, New Zealand and Australia). Through most of human history, the objective of societies was to create stability, because stability meant survival. The tool kits of our ancestors were finely tuned to match the local environment, helping us harvest the food and other resources we needed, but in turn those tools were based on what was available in the location where we lived. One of the major characteristics of modern Western society was the establishment of long-distance trading that allowed the exploitation of resources globally and thus broke the intimate connection between people and their local environment. Although the power of Western culture created the first truly global economy, it was not the first time that cultural expansion was made possible by technological power. The rise of various empires such as those found in Egypt, India and China were based in part on technological developments and long-distance trade. The initial spread of Islam around the globe was another historical period that had a strong technological aspect. Yet in all the earlier cases, a period of invention was followed by a period when invention declined or was actively resisted. Too much change brought about by technology created a social reaction to restore stability.

This examination of the history of technology ends with a focus on Western society because currently the West has produced the greatest abundance of tools, machines and infrastructure. Western powers came to dominate international relations in the last 500 years in large part because technology gave them an advantage over other groups of people. Yet as the technologies pioneered in Western countries became globalized, the advantages have narrowed or disappeared. There is thus a distinction in this book between "Western society" used to refer to the geographic collection of European-based societies and the historical period of Western expansion and colonialism and "industrial society" that includes the West but also other countries that have created significant manufacturing economies such as Japan, Korea, Turkey, China and South Africa.

The rise to power of Western society makes it easy to conflate the idea of technological power with cultural superiority, but this is just another version of chauvinism and no different than claiming that cultures can be ranked by skin color, language or religion. What makes the problem of technology and culture more complicated is that from a historical point of view, and in terms of physical condition, we as people *are* better off than our ancestors. Although the benefits are not uniform, and the disparity between rich and poor cannot be overlooked, it is nonetheless the case that people in the industrialized world (not just the West) are the longest-lived, strongest, smartest, healthiest people in human history.[1] This is not to say that industrial society is without problems (there are many), but rather to point out that the effect of technology can be looked at from different perspectives.

People who object to the idea of technological society often suggest that while people in the industrial world may be better off in a material sense, all this technology has not made us better people. As for being better people, it is difficult to measure whether we are better or worse in a moral sense, but by other measures we seem to be doing very well. We are safer than people in the past as overall levels of violence (outside of actual wars) have steadily declined, the concept of human rights has developed in parallel with the growth of industrialization, and our ability to empathize and offer aid to people around the world has grown with our ability to communicate and travel. Philosophical ideals such as universal suffrage, human rights, public education and democracy have become a reality in parallel with the rise of industrialization. Technology made it possible, and perhaps even a necessity, to care about people beyond the family or the tribe.

It is likely that you have read the above statement with a growing sense of suspicion. Few people when presented with an argument extolling the virtues of technology can avoid thinking of counterexamples to balance the optimism of progressivism. Hasn't our technology also given us the power to destroy all human life on the planet with weapons of mass destruction, and did we not fight the two biggest wars in history where killing was "industrialized"? Our industries and lifestyle are wrecking the environment, depleting the ozone and producing climate change. While we have a powerful medical system, we are also producing new health problems such as obesity and a plague of diabetes, environmental sensitivity and cancer. Some people have even argued that we have enslaved ourselves to the demands of machines, from our Pavlovian response to ringing telephones to our dependence on the complex systems of energy, communication and industry that keep contemporary society going. There is a subculture of survivalists and preppers who are preparing for what they believe is the imminent collapse of industrial society. Popular culture is full of stories about dark futures where monsters of our own creation lurk, from Mary Shelley's *Frankenstein: Or the Modern Prometheus*, the first great morality tale about the dangers of modern technology, to *Gattaca* and the *Terminator* movies.

These reactions to technology are also the product of technology.

This is exactly the challenge that historians face: How do you present the history of technology without falling into the trap of presentism and progressivism, or taking the opposite position and becoming completely anti-technology?

The first answer is that historians try to see all parts of human history as valuable in their own right. We ask about what issues were important to the people of the time and how the people of the past acted to address those issues. This helps historians link the past with the present without making it seem that the past was just getting the world ready for us.

The second strategy is to understand that the level of total complexity in human society remains relatively constant. We falsely believe we live in a more complex society than people in the past because we live in a world filled with complex devices, see things that are happening around the world, and have access to massive quantities of information. While we expend a great deal of time learning how to live in a world filled with devices, from coffee makers to laptop computers, people in the past filled their lives just as fully with the skills of hunting, farming and dealing with the people and spirits that filled the world. Consider television as a form of entertainment. Producing a television program is an incredibly complicated activity, but watching one requires little or no effort except sitting and staring at the screen. For our ancestors, home entertainment required things like a knowledge of how to play an instrument, remembering lyrics, poems and stories, learning the rules for games, and participating in religious and ceremonial rites. All those activities depended on the active participation of the people involved, not just passive observation.

Another aspect of the complexity of the past has to do with memory. People in the past remembered far more than people in literate societies who have transferred memory to paper or digital form. From important events to poetry and music, human memory was just about the only means of recording available to people who did not read or write. It was not uncommon in pre-literate times for people to memorize hundreds and even thousands of lines of poetry, dozens of stories, or the lyrics to a song after hearing it only once. Today such memory work is seen as a special talent or requiring the kind of serious effort a stage actor uses to memorize a part.

In addition to the idea that social complexity is not analogous to the complexity of our tools, we do well to remember that technology is never trapped in amber. We mix tools, systems and approaches to problem solving from across time and from many places. If you open the toolbox of a carpenter today you might find a hammer, a ruler, a spirit level, a screwdriver and a power drill with a lithium ion battery. In one little box we have tools that span at least 5,000 years of technological history and devices created in a variety of different places and cultures. This tangling of technology chronologically and geographically has been looked at by David Egerton, particularly in his *The Shock of the Old: Technology and Global History since 1900* (2007). He argues that technology can appear, disappear and then re-emerge as well as undergoing transformation when it is transferred from one place to another. This allows "old" technology to be mixed with "new" technology as people use a variety to approaches to solve problems.

A final point about the complexity of human life comes from Neil Postman, who pointed out that information is not knowledge. We tend to discount the knowledge of our ancestors because we assume that we know more about the world today than people in the past knew. What this really means is that we have more information about the world and it tends to be more accurate because we have developed precise measuring devices, have access to vast information systems, and use scientific principles like experimentation to gain that information. What people in the past lacked in terms of access to information, they more than made up for with local knowledge. They knew their world, and more, they knew their place in it. Rather than seeing the world in terms of data, they saw things as a series of relationships, often connected to religious or spiritual worlds beyond the immediate physical world. For historians, it is important to be cautious about assuming that our access to information means superior knowledge. A perfect example of this is the completely false idea that people in the past thought the Earth was flat. The majority of scholars from the Babylonians to the scholars of the late Middle Ages thought the world was a sphere. Sailors around the world knew the world was a sphere. Christopher Columbus did not set sail to prove that the world was a sphere, but to find a new route to Asia. The question for geographers of the past was not the shape of the Earth, but its size and whether people lived in other parts of it. If we impose ignorance on our ancestors, our understanding of our own history will be flawed.

⚙ Technology in Society versus Technological Determinism

Societies exist because they are able to exploit their environment to gain the resources necessary for survival. The only way to exploit the environment is by using technology, so societies cannot exist separately from their technology. As the range of tools and number of people has increased over time, the relationship between technology and the people who use it has become more complicated. Since social rules both perpetuate and constrain technology, there is always a tension between the need to use technology and the need to follow the rules about the use of technology. This tension is particularly evident when new tools or methods are introduced that change the relationships of the people within the society and the people and their technologies.

In contrast to this interactive view of technology in society is the unidirectional model of technological determinism. At the most basic level, technological determinism seems perfectly reasonable. For example, humans have dreamt of flying since the dawn of time, but we could not actually undertake controlled flight (not just floating or gliding) until we had developed the internal combustion engine to the point where it could provide the power to propel an airplane. It would then seem perfectly logical to say that human flight was dependent on the availability of the technology of the internal combustion engine. It follows that the modern aviation industry, from fighter jets to package tours to exotic destinations, was only made possible by the existence of a specific technology.

One of the earliest applications of technological determinism to explain history was by Karl Marx. His most famous assertion about the relationship between devices and social organization was in *The Poverty of Philosophy* where he said "The handmill gives you society with the feudal lord; the steam-mill, society with the industrial capitalist." By explicitly linking a certain tool with the structure of society, Marx was pointing out that history depended upon the material conditions experienced by people.

This idea has often been repeated, although Marx's ideas about technology were more directly concerned with the means and control of production than with the equipment being used. In 1967, the historian Robert L. Heilbroner addressed the concept of technological determinism by asking the question "Do Machines Make History?" in an important article of the same name (Heilbroner 1967). His answer was almost as complex as the issue itself, but concluded in part that the degree of effect of technology on society depends on the state of the society at the time of the introduction of the technology. Thus, the greatest degree of technological determination occurred when capitalism was least restrained: "Technological determinism is thus peculiarly of a certain historical epoch … in which the forces of technical change have been unleashed, but when the agencies for the control or guidance of technology are still rudimentary."

When we think of technological determinism in terms of what we can do in the physical world, it seems perfectly reasonable. If we apply the theory of technological determinism to how we behave, or how we interact with each other, it fails.

One of the most profound applications of the concept of technological determinism has been the idea that the rise of modern mass democracy was dependent on the development of mass communications, specifically the invention of movable type printing by Johann Gutenberg around 1450. The reasoning for this conclusion hinges on the idea that, to create a democracy, the people must be aware of the issues and be able to discuss, plan and report on actions. Candidates must be able to communicate with the electorate. Without the ability to communicate with a significant portion of the population, no such coordination would be possible. Prior to mass printing, democracy could only function in groups small enough for the voters to attend meetings or have personal knowledge of the issues and candidates because that was the only way to get the information necessary to participate. Thus, democracy could work in a limited way in a city-state like Athens, but not for a large country.

In the determinist story, mass printing solved the communications problem, and candidates no longer had to personally interact with voters to gain their support. It also created higher levels of literacy, increased the speed at which ideas could be communicated, and raised the expectation of the people in regards to their ability to participate. It is certainly true that the modern democratic states only came into existence after Gutenberg's printing press had been spread across Europe and into the Americas.

There is a flaw in this argument. If there was strict technological determinism, it should follow that printing leads inevitably to democracy, but of course this is not the case. Some of the most powerful totalitarian regimes came into being *after* printing was introduced, and they used mass media to create and maintain control of their people. Totalitarian governments continue to exist even with the addition of electronic media to the print world. Noam Chomsky in *Manufacturing Consent: The Political Economy of Mass Media* (1989 and 2000) made a powerful argument that, even in democratic states, mass media (starting with print) are used by the powerful to control society for the benefit of the powerful rather than to further democracy. So, from a historical point of view, we can say that print may have made mass democracy possible, but democracy was not an inevitable product of printing. Gutenberg's invention did not, simply as a device, give us modern democracy, but in some societies the capability of print was used by people to pursue their political objectives of democratic reform.

Another problem with strict technological determinism is its assumption of an inherent drive for efficiency and utility imbedded in technology *and* the technology-using society. Although most technological determinists are not teleologists in that they do not believe that there is a specific end purpose to history, there is often an underlying belief that evolving systems work toward some form of perfection. This idea is linked to a view of technology that assumes that it is in some sense natural and follows a kind of evolutionary paradigm, in which technology evolves from simpler to more complex forms. This model also suggests that technology changes so as to maximize our control over the physical environment by perfecting our knowledge of nature (thus science can be seen as serving technology) and then turning that knowledge into systems to extract what we want from nature.

While it is certainly the case that in the long term our technology has become more efficient and complex, the determinist model does not account for the periods of human history when people have resisted technological change, and in fact tends to portray such periods as anomalous or times of decline. Thus, eras of technological stability are equated with some kind of failure, and even as unnatural. Determinism thus assumes that people are passive consumers of technology, which is historically incorrect. Determinism also assumes that materialism is the best and perhaps only legitimate way of judging whether a society is a success, and this is also a highly debatable assumption.

One of the most important thinkers to address this issue was Jacques Ellul (1912–94). In his book *The Technological Society* (1964), he introduced the term "technique" to separate the artifact from the system. He defined technique as "the totality of methods rationally arrived at and having absolute efficiency (for a given stage of development) in every field of human activity" (Ellul 1964: xxv). Such efficiency came at the cost of our concern for each other. Ellul, who was a Christian theologian, was very concerned that the drive to perfect technique gave us increasingly powerful devices and systems, but at the cost of our humanity and perhaps even at the cost of our souls. Although his position assumes a large degree of technological determinism, his idea that technology has a moral context is important to remember.

⚙ Other Views of Technology: A House to Live in or a New World?

The ideas of two specific thinkers have helped shape the philosophical argument of this book: Ursula Franklin (1921–2016), scientist and philosopher; and Neil Postman (1931–2003), scholar of education and media critic.

In 1989, Franklin delivered a six-part radio talk for the Massey Lecture series entitled "The Real World of Technology" (Franklin 1999). The Massey lectures ask leading intellectuals in diverse fields to speak to the public about their field of research. Franklin, who was a pioneer in the study of materials science, was deeply troubled about the way that technology was changing human life. She was concerned that if we saw technology and society as the same thing, that is, if we were technological determinists, we effectively gave up our ability and right to control technology. She started her lecture by using an analogy, arguing that technology is the house that we live in. Over time, more and more of our activities take place within the house, until for many of us the house is the only world we know and we increasingly lose our awareness of anything outside it. For Franklin, this meant that technology was necessary for human survival, but was separate from people. The "house" is under constant construction as new parts are added and old parts remodeled, but so long as we understand that it is separate from us, we have the ability to control the building process.

Part of the problem about controlling technology, according to Franklin, was that our perceptions of the world around us were shaped by the techniques (in the sense that Jacques Ellul defined the term) we employed to interact with the world. For example, an engineer designing oil drilling platforms might speak of the dangers of "ice-infested waters" while an environmentalist might speak of the dangers of an "oil rig infested ocean." Neither perspective is right or wrong in some transcendental sense, but they represent potentially incommensurable views. If the engineer and the environmentalist can't understand each other, the resolution of problems becomes very difficult. Franklin reminds us that if we lose sight of the fact that technology is constructed, we will assume that it is inevitable and is therefore in some way the correct or natural state. If certain technologies come to be seen as inevitable, then other perspectives and possible methods of solving problems are precluded. For Franklin, there was an "outside" to the house of technology that human society had constructed. The outside was nature, and since humans were part of nature, to balance and keep technology in proper perspective, humans must remember and cherish the natural world.

Although Franklin views technology as having a greater degree of separation from society than I think it has, her concern about what might be called technological blindness is an important idea. If we treat technology as having an autonomous existence rather than being an interdependent component of society, we will create social relations based on the belief that technology is beyond social control. This would have dangerous repercussions, not the least of which is a sense of helplessness when dealing with technology.

A different perspective on the distinction between technology and society comes from Postman, particularly from his book *Technopoly: The Surrender of Culture to Technology* (1992). Postman discussed the importance of technologies such as language and education, what are called "invisible technology" that does not produce a material artifact, and I use this idea extensively in this book.

Like Heilbroner, Postman believed that the degree to which society is shaped by technology depends on what period in history you look at. In the far past, tools were personal devices and made human life easier. Technology in that early time simply extended our natural abilities. We made the tools and were in charge of the technology. With the Industrial Revolution, technology gained a certain level of autonomy from social control and new machines became too complex for construction or control except by people with specialized knowledge and skills. A new class of technicians such as mechanics and engineers were needed to create and service the new devices. We were still largely in control of technology, since it was still based on mechanical principles that could be worked out by observation and, although much more powerful than hand tools, was still just an extension of muscle power and manual dexterity. With the introduction of electricity and the computer age, Postman argued, technology has gained almost complete autonomy and we are now forced to change society and even our personal relationships to suit it. Working with the new technology required a great deal of training, and some systems have become so complex that no single person can understand all of it. The way devices work is not obvious, or even observable. The easiest way to understand this problem is to look at a waterwheel and the inside of a computer. It is self-evident how the waterwheel works in the sense that a person can, by observation, figure out what all the component parts of the waterwheel are doing. It is not self-evident how a computer works, nor could a person without special training even determine what role the component parts play in its operation. This last era he called "technopoly" and he warned us that in a technopoly our devices and systems have gained autonomy over society and reduced our self-determination and narrowed human relations.

For Postman, technology is not additive, but transformative. We are not adding new rooms to the house when we introduce a new technology, but transforming the whole system. In other words, it isn't society plus the automobile, it is a new society. To understand the effect of technology therefore requires an understanding of society that includes such things as the dominant philosophical ideas of the age (conservative or liberal; expansionary or inward looking); the structure of the society into which the technology is introduced (caste system; theocracy; democracy); and the economics of the society (barter system; feudal dues; laissez-faire capitalism). In other words, if technology is a system and not just the material end product, then understanding why technology emerges at particular times and places, and the changes that come from new technology, requires understanding the interaction of systems, not just the immediate conditions associated with the new invention.

While Postman offers some powerful insights, he underestimates society's ability to both adopt and adapt technology even in an age of specialists and complex systems. For example, no one at the dawn of computing would have predicted that social networking would be one of the most powerful uses for computers. We may not understand what is inside the computer, but collectively we seem to understand that computers can be made to solve problems.

✿ Other Voices

Good history is not just a record of success. It looks at struggle, conflict and failure and at the role of a range of people who helped shape events and were in turn affected by those events. This broad-based social and cultural history has its roots in the philosophical ideas of the Annales School and *nouvelle histoire* or the new history movement. Although there is not a unified philosophical code for this historical approach, social and cultural historians who ascribe to this way of looking at history argue that history is not just made by "great men" such as kings, generals and popes, but must reflect the reality of the populace. In the earliest prehistoric period before written records, inventions appeared without a known inventor and it is easy to look at the social and cultural utility of devices. As we get nearer to the present, it is tempting to see history of technology as a series of successes created by particular people. Although it is important to give credit to the creative people who helped shape our society with their inventions, it is easy to let the inventions overshadow the larger picture.

One of the ways to try and balance the problem of following innovation with social effects is to recognize that there are other voices that have traditionally been underrepresented in the pages of history. In particular, women, non-Europeans and indigenous people have, until recently, been almost completely invisible in accounts of Western history. The problem of finding and representing the other voices of history poses a particular challenge for a history of technology. While non-Europeans made up the bulk of the early history of technology, with the start of the Industrial Revolution, European interests and later American developments came to dominate. While it would be disingenuous to say that technology was not predominantly westernized in this period, the spread of technology was not uniform or without conflict. Those conflicts are as much a part of the history of technology as the innovations.

The role of women or indigenous people in history can be even harder to elucidate because there are often few if any records. This leaves historians trying to infer the history. Consider the problem of women inventors. From the position of individual psychology, women and men are equally curious about the world and equally inventive. Indeed, some archeologists argue that women were probably the first primary tool users. Observations of chimpanzees indicate that females use tools to extract resources more than males, and female offspring spend more time observing their mothers than males. There is no reason to believe that stone tools or the controlled fire were discovered only by men. Yet female innovators are hard to find in the historical record because men came to dominate the public sphere and inventions are propagated through the public sphere. Moreover, men were responsible for the production of most historical records, and in patriarchal societies men control all resources including the intellectual property of female dependents. As job specialization increased over time, women and indigenous people were prevented from becoming mechanics, engineers or technicians and therefore prevented from gaining the skills, connections and access to financing necessary to turn ideas into marketable products. In heavy industry, women were excluded because of ideas about protecting women from danger and physical exertion. These excuses continued to be used long after mechanization and safety standards had transformed the shop floor. Today, in Silicon Valley (an area where the hardest physical activity is carrying a laptop and a cup of coffee) women are actively or passively excluded from positions of power and authority and thus their role in the invention of the computer age is hidden.

Part of the answer to this problem is to include under-represented people whenever they can be, so that the stories of people like Grace Hopper, one of the pioneers of computing, become part of the history. Another way to offer some balance is to recognize that while particular groups have controlled the propagation of technology, everyone is affected by technology, and the history of technology is about more than the invention of new artifacts. It must include the effects, both positive and negative, of the transformations caused by the introduction and use of technology.

🍥 Conditions for Technological Change

Historians of technology have been very interested in what conditions must exist for technology to be developed and accepted by society. Sometimes it seems to evolve over years or even generations, but at other times the emergence of a new device or tool seems almost instantaneous. In world history, there are clearly times when invention and innovation are common, and then other periods when few or no new technological ideas are pursued. Different regions of the globe also have differing levels of acceptance of technological innovation that can also change over time. What seems to be the case is that there are conditions favorable to innovation, but there is no way to insure that innovation *will* occur, or that no innovation happens at times when the conditions are less favorable. There are three contingent conditions for periods of high rates of innovation:

1 competition
2 a cultural attitude favorable to novelty
3 social flexibility.

Competition

Competition can be of the kind that occurred in the Arms Race when weapons development during the Cold War spurred a great deal of innovation, but historically war is not actually the best form of competition for innovation. What tends to happen in times of war is that most innovation is directed toward the perfection of existing weapons, for the simple reason that the combatants cannot afford to invest much time, money or intellectual resources on devices that might not work. A good example of this is the appearance of aircraft in the First World War. The use of fixed-wing aircraft was a significant aspect of the war and there were innovations made during the war such as fuselage-mounted machine guns and the bomber, but there was far more work on military aircraft in the years following the war.

The kind of competition that seems to foster the most innovation might be called a marketplace of ideas within a common cultural context. Some examples of where and when this condition has existed would be the Greek city-states around the time of Plato, the empires of the Islamic Golden Age, and the corporate system in Britain during the Industrial Revolution. Each of these eras is notable for the blossoming of culture, scholarship and technology. In each case, the innovations took place in a region that shared cultural commonality, even if there was not political unity. Having a common language, for example, made the transmission of ideas easier and faster.

A Cultural Attitude Favorable to Novelty

Discussing the zeitgeist or spirit of an age is always a tricky proposition for historians, but it is the case that societies that seem to be more interested in new things favor technological innovation. During the Victorian era in Britain and western Europe, there was a huge appetite for novelty, and both the wealthy and the middle class had cabinets of curiosities, often containing displays such as exotic stuffed birds from faraway lands. At the same time, organizations such as the Mechanics' Institutes were set up to educate working people and present lectures and demonstrations of the newest ideas in science to the public. Starting in the 1840s, France, Britain and the United States all held wildly popular exhibitions highlighting technology. Everything from automated weaving machines to street lighting to the kitchen of the future was shown to the public at such exhibitions. The peak of the Industrial Revolution overlapped with the public's fascination with new things.

Cultures that don't show much interest in new things seem to resist innovation. The turning inward of China after the voyages of Zheng He and the end of the Islamic Golden Age were periods when innovation in those regions declined. Sometimes innovation declines because the economy falls, or incessant political turmoil and warfare disrupt the society so much that there simply is no market for new things. For reasons that are even less clear, sometimes societies seem to say "we have had enough change." The futurist Alvin Toffler (1928-2016) coined the term "future shock" to put a label on the psychological impact of constant and rapid technological change. While the focus of his book *Future Shock* (1970) was on the potential damage to the individual brought about by rapid change and information overload, such ideas seem equally applicable to societies.

Social Flexibility

Social flexibility can mean that there are not rigid classes in a society and social status is determined by some measure of merit, but it is not the case that highly class-based societies with low social mobility are inherently less innovative than egalitarian societies. An educated elite, such as China's senior bureaucrats or Egypt's priests, can be very inventive, especially if they are responsible for dealing with engineering problems. In a more general sense, what social flexibility means is that people are not prevented from innovating simply by their class. It also suggests that the people who come up with new ideas can directly benefit from their work. The farmers of the early Islamic world introduced many important innovations, but largely remained farmers. Such innovations did mean that a greater portion of the population did not have to stay on the farm, so the sons of the farmers could become scholars, artisans or merchants. In the longer term, social flexibility means that the society can adapt to the changes that new technology triggers. Societies with greater social flexibility tend to experience more innovation.

Similarly, there are conditions that are associated with low levels of innovation, such as high levels of social turmoil from economic collapse or war. Regimes that enforce a strong social ethic of conformity often have low levels of innovation since innovation may threaten either the authority of the rulers or the general stability of the society. Low levels of social flexibility are also often associated with low levels of education. A lack of an educated populace, particularly in the post-Industrial Revolution era, can be linked to a lack of technological development. This is not to say that poorly educated people are not inventive; they frequently need to be very inventive to overcome daily problems. What happens is that local innovation is slow to spread and can be difficult to develop into commercial ventures.

A related issue is whether innovation can be made to order. In the industrialized world, governments have been so concerned about innovation over the last 300 years that they have tried to foster technological research through support for things like education, the offer of prizes (such as the prize offered by the British government for a way to measure longitude as sea), grants and loans for research, the creation of "centers of innovation" or tax-breaks for start-up companies. Governments also engage in direct investment. In some cases these strategies have worked, so that the prize offered to find ways to calculate longitude at sea really did lead to serious and successful efforts to solve the real-world problem. The Russian czars had less success in their efforts to transplant technological systems by importing experts and setting up institutions based on models they found in France, Germany and Britain. The steam locomotive, the aerospace industry, computers, cell phones and genetically modified crops are examples of modern innovations that have had major government support, but such support is no guarantee that a technological breakthrough will follow or that society will readily accept the new technology. The money wasted on trying to develop X-ray lasers for the Strategic Defense Initiative (the so-called "Star Wars" defense system) is one of the best examples of the difficulty of trying to command new technology to be created. While direct attempts to stimulate innovation have had mixed results, whenever societies have created a class of technologists, whether in the form of the mechanics of the Renaissance or the engineers of the École Polytechnique founded in 1794, there has generally been a rise in innovation and a greater reliance on tools for solutions to problems that the society faces.

✿ Winners and Losers

One thing that almost all commentators on technology agree on is that the introduction of any new technology results in winners and losers (or those who manage to benefit or exploit a new technology and those who do not). The simplest part of this has to do with employment, so that people who made bows and arrows were not needed by soldiers armed with muskets. When printing became widespread, scribes were less in demand, and more recently the vast numbers of secretaries who made up the typing pools of large companies disappeared when the desktop computer invaded the business world. Sometimes the stories of employment displacement can be terrible, as suddenly people who were respected and contributing members of society find themselves destitute, as weavers did after the introduction of power looms. In some cases, the individuals displaced by new technology were absorbed into other jobs, but often they fell into poverty and unemployment.

Painful as the experience of the individual losers can be, new technologies have tended to create more jobs than they eliminated. While a farmer with a plow could do the work of three or four farmers without plows, the surplus from plow-based agriculture could support at least that many people who then did not need to be working the land. For every weaver displaced by a steam-powered loom, there were new jobs created in manufacturing, not to mention transportation, sales, advertising and management. For someone strongly pro-technology, the short-term pain of the displaced workers would be vastly outweighed by the long-term gain for society and future workers.

The winners in the technological world tend to be those people who become wealthy because they are in a position to utilize the new technology. Many mill owners during the Industrial Revolution gained wealth equal to the richest monarchs of the ancient world, while today there are thousands of hi-tech millionaires. New classes of employment, often highly technical and very well paid, have been created with the introduction of new technology.

There is a deeper level to the issue of winners and losers, however. In a larger sense, new technology has created winner and loser regions and cultures. Colonialism was driven in part by the demands for cheap raw materials to feed the growing industrial economies of western Europe. The consequences of slavery and the displacement of local populations are still with us from that time. It has not just been jobs that have disappeared because of technological change. Cultures, in part or in total, have been wiped out by it. Military might has been used to seize land and enslave people, and commit genocide, destroying communities and culture in the most direct fashion possible. In a more subtle fashion, mass communication, for example, has contributed to the disappearance of hundreds, if not thousands, of languages around the world. The ties of work, faith, art and family that have bound communities together have often disappeared as mass production and mass media provide an attractive and inexpensive alternative to the old ways. Those people with the most powerful technology can either deliberately or indirectly impose their culture on those around them. Just as losing genetic diversity from food crops poses potential problems, so does the disappearance of cultural diversity.

In a sense, any history of technology is a history told by the winners, since those societies that failed to understand the problems of their technologies, or who were unable to overcome real-world problems with workable solutions (whether technological or otherwise) get erased from history, leaving only archeological remains. Our current global society is not exempt from this condition.

✿ Technology Traps

Technology offers us the ability to do things that we could not otherwise do, whether it is bringing down a mastodon with flint spears or flying to distant vacation spots. Yet technology requires things from us as well. When humans started to domesticate plants and animals, we were also domesticating ourselves. We gave up the autonomy and flexibility of the hunter-gatherers or foragers for the stability of the farmer. When we built modern cities around the automobile as the main mode of transportation, we locked ourselves into communities that are not designed for human-scale or human-powered activity.

The foundation of technological traps is that the perfection of certain devices and techniques seems to follow logically from the original intention of the invention, but leads to unexpected and negative consequences. For example, during Paleolithic times, the development of better and better stone points for spears, arrowheads and knives meant that no big game was beyond hunting. As a consequence, many of the large animals were hunted to extinction. In places, the hunters literally denuded the landscape of big game, then were forced to move or starve. Humans may have spread across the planet because they feasted until the cupboards were bare and they had to move on. When we learned to farm, a number of civilizations rose to take advantage of a brilliant invention – irrigation. This led to a sudden rise in agricultural production, followed by a rise in population, but it was a trap because irrigation deposits salts on the fields. Without careful management, the salts turn the fields into lifeless sand, followed by agricultural and social collapse. The only solution that some civilizations could come up with was to introduce more irrigation and irrigate more land, putting off the collapse for a time, but in the long run making the damage even worse.

A similar story can be told about wood. Almost all early civilizations required large quantities of wood for construction, but, more importantly, for fuel. When the local trees were gone, some system of transport had to be started to bring wood to the settlement, but without some management of the resource, all the wood that could be economically transported was used up. The most extreme example of this comes from Rapa Nui or Easter Island. Once a tropical paradise covered with trees, when people arrived they eventually denuded the island. In the small space, it must have been apparent that eventually all the trees would be gone, but the people still cut down every single tree, leading to a major social collapse. A number of ancient civilizations, from the Americas to Asia, rose and disappeared because the people became so adept at using resources (that is, they perfected their resource extracting techniques) that their populations rose above the level that could be supported by the land within the range of control and use of the city's people. Sometimes this meant the collapse and disappearance of the society. Other times, it led to territorial expansion and the creation of empires, but in a sense that could only put off the problem if the rate of resource extraction continued to be greater than could be renewed by the supporting environment.

Today, the use of petroleum is presenting us a perfect technological trap. So much of modern industry depends on hydrocarbon products, not just as fuel, but for lubricants, plastics, and the base stock for tens of thousands of products. Peak oil is not just about the rising cost of fueling our automobiles, but about making every aspect of modern life possible. We have known for generations that the oil would run out, but we have invested so much in petro-culture, that changing our society will be very difficult. We have created almost perfect machines for the use of petroleum products and then produced billions of them without control or thought of the future. Will we be like the Easter Islanders and use up the last drop without planning what we will do next?

The "Dollar Auction" and the Price of Winning

In 1971, the economist Martin Shubik published a paper outlining a game that illustrated why people could get trapped into overspending in a competition (see Shubik 1971). The Dollar Auction Game was simple. A dollar was put up for auction with 5 cent increment bids. There were only two rules: (1) highest bid won; (2) all other bids were forfeited.

At first, it seems like easy money. A bid of 5 cents could return a profit of 95 cents. As bidding went up, the potential reward went down until it became apparent that the second rule really determined how people would act. If Person A bid 90 cents and lost to Person B who bid $1 (neither losing or gaining anything), Person A would lose 90 cents. To minimize loss, Person A would bid $1.05 and lose only 5 cents. In real trials of the game, people consistently bid over the dollar, sometimes bidding 10 dollars to win the dollar. In fact, the only winning strategy is not to bid.

What Shubik and others pointed out was that the Dollar Auction forced people to rationalize overspending in terms of potential loss rather than potential gain. This idea shows up in the real world in military arms races (dreadnoughts and nuclear weapons), and political struggles (the Cold War). It also has implications for the adoption and spread of technologies. Take, as an example, a competition between two cable television providers: Alpha Communications and Omega Cable. To be profitable, a provider needs 60 percent of the market. Since the equipment available to each company is the same and the service being offered to the customer is identical, there are few ways to gain market share, but a stalemate at 50 percent of the market for each company means they both lose money. The only option is to offer the service at a lower price, leading to a price war that sees each company offering services at below the actual cost of providing the service (in effect, bidding more than a dollar for a dollar). Omega Cable has investors with more money so it wins, while Alpha Communications goes bankrupt, wiping out its investment. With 100 percent of the market, Omega has little reason to innovate and can raise prices to be very profitable. Actual examples of this kind of battle range from the telecommunications industry to supermarkets.

The Tragedy of the Commons

Another idea that has implications for the impact of technology on society is the tragedy of the commons. The term comes from a pamphlet on economics by William Forster Lloyd (see Lloyd 1833). Lloyd pointed out that cattle herders often had their cattle graze on shared or common land. If some of the herders were selfish and let more cattle graze on the common than the land could support, in the short term they would benefit, but in the long term everyone would lose. In 1969, the biologist Garrett Hardin used Lloyd's observation to point out that the same form of problem can be found in the control of pollution, attempts to legislate temperance, the nuclear arms race and most importantly, overpopulation (Hardin 1969). Today we could add climate change, the drug trade and internet neutrality as modern examples of the struggle over what things are held in common.

Essentially, Hardin argued that the freedom and logic of self-interest doomed people to destroy nature and society. The only solution was to use "mutual coercion," that is, we would mutually agree to limit our own freedom for the collective good. Hardin was attacked on a number of grounds for his argument, but the central idea that nature is limited (in this sense meaning natural resources) but demand grows with population remains powerful. Technology plays a complex role in the tragedy of the commons. On one hand, the industrialization of the world has increased the speed of resource consumption and increased pollution. On the other hand, we have often found technological solutions to problems that have decreased our impact on the natural world. London no longer has killer fog, and alternative energy (although not completely without impact) is steadily growing in use and utility.

Technology, Networks and Communication

One of the recurring themes of this study of technology is communication. Technology never exists separate from the society it was created in, but how technology is spread, used, and viewed by the people of the time depends a great deal on the ways information flows. The faster information moves, the more likely technological change will also go quickly. In the modern world, we are so conditioned to high-speed information systems that news that is just twenty-four hours old is considered almost not worth printing or broadcasting. The systems we use to move information, from the invention of writing to the internet, has been one of the most significant areas of technological development, and the various inventions have had world-changing consequences.

Important as the various technologies to communicate have been, there is more to communication than simply moving information around. Historians, philosophers and, more recently, media theorists have tried to understand how the means of communication affects our understanding of the world. The most famous of these theorists was Marshall McLuhan (1911–80), who made the study of mass media a mass media subject. He pointed out, in a pithy quote in his book *Understanding Media: The Extensions of Man* (1964), that "the medium was the message." McLuhan's ideas about media were complex, but part of his point was that the way information was communicated shaped our response to it. There also seemed to be a historical link between the means of communication and historical change, so that the Protestant Reformation and the French Revolution were, in part, made possible because mass media (books, pamphlets, posters and newspapers) was made possible by the print revolution.

In addition to McLuhan's ideas about media, the scholar Harold Innis (1894–1952) wrote about the importance of communication in the creation of civilization, especially the formation of empires. In his book *Empire and Communications* (1950), he traced how the relationship between the means of communication (such as oral, stone tablets or paper) available to civilizations has helped shape those empires. One of his most profound observations was that the River Nile was not just a transportation system, but a communication system. Whoever controlled the Nile, controlled the flow of information in the empire, and thus controlled the empire not simply by law or force of arms, but by determining what people knew and thought about.

In terms of the history of technology, many of the most important discoveries and inventions have made it possible to create new forms of networks. Roads in the Roman Empire, the telegraph in the Industrial Revolution, or the electromagnetic spectrum of radio, television and cellular phones in the modern era are all examples of networks that allowed the larger social structure to come into being. If Innis and McLuhan are correct, then the shape of the society was, at least in part, the result of the type of communication network the technologies of the day made available.

Although there are dozens of forms of networks, many of the most common look the same when broken down into their component parts. A feudal hierarchy and an electrical distribution system share a number of common factors (Figure 1.1). Each circle represents an information-generating node with formal or direct links with other nodes. In the case of the feudal system, those connections are based on social relations, while in the electrical grid they are the wires and the electrical demand. There are also informal or indirect information paths which may pass in and out of existence. A monarch who fails to pay attention to what is happening among the serfs can get into trouble, but only hearing about serfs through the filter of the formal paths could distort the real story, so a monarch might use spies or informal contacts outside the regular social structure, or even visit people in disguise.

For the electrical system, power demand is an almost perfect form of information transfer. As demand rises in the morning when businesses, schools and shops start opening, the demand for power is completely known to the operator of the generator,

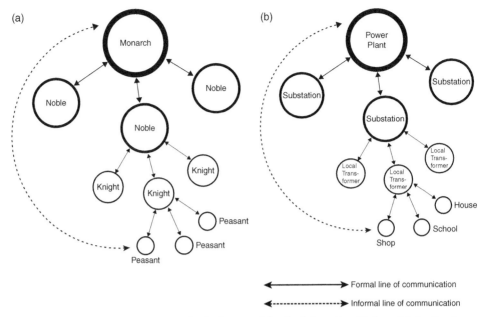

Figure 1.1 Two communication networks. Each network has both formal and informal links. For the monarch, formal links are the court officials, but to avoid isolation a monarch may have meetings with commoners or use spies. In the power system, formal links are determined by the power grid where an information link is created by monitoring demand to increase or decrease production of electricity. An electrical producer's informal link could be discussions with a city council about city growth leading to higher demands for electricity.

who then increases output to meet the demand. A wise power operator will want to keep in touch with the end users to determine if there are problems, plan for the expansion of business or even keep track of weather.

✿ A Final Word

This history of technology is meant to be not an exhaustive study of every major innovation, but rather an examination of how technology has shaped human history. In some cases, the text looks at the life of certain inventors whose impact on history has been as significant as the devices they made. In other cases, how a device works is highlighted because understanding the "nuts and bolts" (a great technological phrase) offers a useful insight into the historical place of an invention. The overarching argument of this book is that civilization and technology are intertwined, with technology transforming society, and at the same time society creating circumstances that lead to the development of new technology.

1 Education and government are types of "invisible technology." What are other examples of technologies that are not based on physical objects?

2 To what degree do you believe that technology determines the course of events in history?

3 Do you agree with Ursula Franklin who argues that technology is like a house we live in or with Neil Postman who says that there is no distinction between technology and society?

NOTE

1. People often accept that longevity and health are indicators of the benefits of industrial technology, but question the idea that intelligence is a product of that system. Yet brain development is highly influenced by childhood nutrition and freedom from illness. Healthy children have healthy brains.

FURTHER READING

The literature on the philosophy and historiography of technology is large and growing. Although this book looks at the history of technology, it draws on a number of philosophical ideas about the nature of technology. For an introduction to the philosophy of technology, the section "Philosophy of Technology" by Maarten Franssen, Gert-Jan Lokhorst and Ibo van de Poel in the *Stanford Encyclopedia of Philosophy* is a good place to start. A more specific discussion of the current state of philosophy of technology can be found in Don Ihde's article "Has the Philosophy of Technology Arrived? A State-of-the-Art Review" (2004). Every historian of technology knows Robert L. Heilbroner's 1967 article "Do Machines Make History?" This foundational article analyzes Karl Marx's influential ideas about the role of technology in society and presents technological determinism in a way that might be useful for historians. Similarly, Lewis Mumford's important work *Technics and Civilization* (1963) remains a staple of history of technology reading lists. Two differing views of technology's place in society can be found in Ursula M. Franklin's Massey Lecture series, published as *The Real World of Technology* (1999) and Neil Postman's *Technopoly* (1993). Both books present definitions of technology and discuss the benefits and dangers of living in a technological society. An important addition to the discussion of how technology functions in society that points out the difference between the use of technology and the intentions of inventors comes from David Edgerton's *The Shock of the Old* (2007). His "use-centered" approach links the systems in which artifacts are embedded with ideas about how we integrate existing ("old") technology with new technology and challenges the idea of simple progressivism in technology.

2 Technology and Our Ancient Ancestors

~ 3.3 to 2.7 million years ago	Stone tools
~ 1.5 million to 700,000	Controlled use of fire
400,000	Spear
300,000	First *Homo sapiens*
300,000 to 200,000	Figurine images of people
170,000	Clothing
110,000	Beginning of Pleistocene glacial period
75,000	Jewelry
32,000 to 29,000	Cave art
25,000	Decorative art
15,000	End of glacial period
14,000	Agriculture
10,000	Ceramics
8,000	Bow and arrow

Our knowledge of our ancient ancestors is founded primarily on the artifacts that survived for us to find. This record of bones, stone tools, cave paintings and pottery tells a story about creativity and society that goes beyond the tools themselves and gives us a picture of the intellectual capacity of our ancestors and their social relationships. Tools extended human physical capabilities, but the technology of early tool use exists only in the context of human society. Planning, education, awareness of materials, experimentation and shared experience are the fabric of the technology that makes a flint axe possible. One of the greatest turning points in technological history was the controlled use of fire, which changed our ability to manipulate matter, gave us a different sense of time and contributed to a physical change in the human body.

There is always a danger when discussing the earliest developments in technology that any observation will be made obsolete by a new discovery. Archeologists and anthropologists continue to build the picture of the lives of our ancient ancestors, and each discovery makes our common history richer and more robust. For example, recent discoveries have pushed back the date for the appearance of tool use among our hominin relatives, have found entirely new settlements, and with the help of DNA evidence have started to create a much more detailed chronological and geographic map of the spread of humans across the globe. As work continues, we will know more about the distant past in the future than we do today. Thus, the historian must write with a certain caution, recognizing that a discovery made next week or next year could radically change our understanding of the past.

Although the details will likely change as research progresses, some aspects of our understanding of the past have only been reinforced by recent discoveries. Technology, particularly in the form of early tool use, was a key part of the lives of our ancient ancestors. It was so important that for many years hominins, and in particular the *Homo* genus running from *Homo habilis* (whose name means 'skillful person' or 'handy man') to *Homo sapiens* ('wise person'), were thought to be unique in having the ability to make and use tools. Even when it was demonstrated that other animals such as crows, chimpanzees and sea otters used implements to accomplish tasks, researchers held that the distinction was not just the use of tools, but the creation of them. In other words, it might be the case that sea otters could find and use flat rocks to open shells or a crow might use a long thorn to spear a grub, but only humans transformed materials to suit their needs. As more examples of non-human tool use were discovered, it became clear that humans were not unique, but rather an extreme example from one end of a spectrum of tool use found in nature. What is true is that most animals that use tools could exist without them, whereas most humans would die without our tools to keep us alive.

Our dependence on tools makes the issue of the origin of tool use a critical question for research on human development. Compared to most animals, humans are ill-equipped to face the rigors of the wilds. We lack large teeth, sharp claws or a tough hide.

Our eyesight is good, but our hearing and sense of smell are poor compared to most other mammals. Without tools, humans would be ranked as "tasty snack" by most large carnivores. Not only do tools represent a constant and necessary heritage that links the present with the past and helps to define what it means to be human, they represent the survival of the species.

In addition to basic survival, tool use is associated with human evolution. Our ancient hominin ancestors used a small range of tools to overcome their physical limitations, but there seems to be a connection between more complex and a greater variety of tools and the intellectual capacity of our hominin relatives. In fact, one of the great questions in archeology has been "did bigger brains lead to tool use, or did tool use lead to bigger brains?" It may be that there was co-evolution, but the archeological record suggests that there were periods of sudden change in tool design and use, suggesting some trigger.

There is a long and sometimes acrimonious debate about the occurrence of the earliest tools and what those tools tell us about our ancient hominin ancestors and us. We will likely never know precisely when or where our ancestors picked up a bit of stone and turned it into a tool. There was probably a very long period when tool use was opportunistic; when the concept of taking a rock or a stick to accomplish a task was understood, but specifically creating a tool and carrying it around for later use was not done. Tool use thus likely proceeded tool manufacturing by a considerable time, but the earliest clearly manufactured stone tools so far identified and dated are between 2.5 and 2.7 million years old. Some scientists have suggested that there are stone tools 3 or even 4 million years old, but these claims are still being investigated.

Not surprisingly, the oldest tools come from the same places that have produced the oldest fossil remains of hominins. The oldest so far found come from the Great Rift Valley region in east Africa. The valley runs from the Gulf of Aden in the north through Ethiopia, Kenya and Tanzania, as well as touching other countries such as Uganda, the Congo and Zambia and ending in the south in Mozambique. Created by the separation of the Nubian African tectonic plate from the Somalian plate, the valley continues to slowly drop in elevation, and in geological time will eventually become submerged, forming a gulf between the plates. Although it is not completely certain that the Great Rift Valley was the birthplace of our hominin ancestors, in ancient times the valley provided abundant natural resources and a temperate climate favorable to hominin existence. The geological conditions also led to the preservation of fossils and artifacts for us to find.

✿ Archeological Eras

Archeologists have traditionally divided prehistoric time into a number of eras such as the Paleolithic and the Neolithic (Table Intro.1).

Table Intro.1 Human Stone Age eras.

Paleolithic	Old Stone Age	~ 2.7 million to ~20,000 BCE (Before Current Era)
Mesolithic	Middle Stone Age	~20,000 to ~8000 BCE
Neolithic	New Stone Age	~8000 to ~3500 BCE

These are based on the style of stone tools found in each period and are thus techno-logically determined, but there is huge debate about both the placement of the divisions and the tool-types that indicate the transition from period to period. The classification system is also somewhat problematic since changes were not geographically uniform, so that the Great Pyramid in Egypt was being built by people who could be classified as being Neolithic or Early Bronze Age, while at the same time the builders of Stonehenge in England were technologically Mesolithic people. Even very recently, archeological discoveries have shifted the earliest dates back thousands of years and future discover-ies may transform our current understanding even further.

Currently the oldest confirmed stone tools were found in the Gona region of Ethiopia by Michael Rogers of Southern Connecticut State University and his team. These have been dated at between 2.5 and 2.7 million years old. They were found in a layer of soil that was covered with volcanic ash above it and magnetite-bearing material below, allowing scientists to use a combination of argon isotope dating (which can be used on volcanic rocks) and magnetic polarity stratigraphy to date the stones. Other tools made of volcanic rock such as obsidian have been dated directly. They come from the period when *Australopithecus afarensis* or another as yet undiscovered contemporary species was thought to have lived, and they are thus our earliest tool-using relatives.

The early stone tools are small, often only a few centimeters in length, and take the form of scrapers, cutters and points. They are different in structure than bits of stone that are produced by rocks being banged together as might be found in a fast-running stream or falling in a rock slide. We know this because we have learned to shape stone (a process known as "knapping") that duplicates the artifacts found in the field. The skill required to make good stone tools is considerable, and experience in both select-ing and shaping stone would have been needed to consistently manufacture good tools. It is likely that many of the tool fragments we have found near the rock sources were the discards, flawed in some way and left behind.

In addition to stone tools, it is likely that other things were used as tools, such as hide, wood or other perishable materials, but no organic material other than bone has survived the millennia to be collected by us. Implements and decorative objects made of harder materials such as bone, shell and teeth have been found in less ancient sites, but it seems reasonable to assume that any hominin who could fashion a stone scraper could use it to sharpen a stick or understand how to dig with a piece of wood or shell. In particular, pointed sticks used as spears are likely even older than the stone tools that have survived. This assumption is partly based on the discovery that some chimpanzees make spears for hunting. The earliest direct evidence of spear use comes from around 400,000 BCE, and spears with fire-hardened wooden tips have survived from 250,000 BCE. Combining the stone point and spear seems to have taken some time to figure out, so stone-tipped spears only appeared around 80,000 years ago.

Stone tools offered our ancestors a pathway to a better life. In particular, the cutters and scrapers made it possible to gain access to more food. Cutting up animals or opening fruit became easier, and at the most basic level that meant higher levels of caloric intake. More food meant more energy, faster growth in children and greater resistance to disease and injury. It may also have contributed to brain development, since one of the major determinants of brain development is childhood diet. In addition to the nutritional benefit, stone tools made it possible to use more of the animal, such as cutting the hide and drilling through bone.

❉ Culture Revealed by Artifacts

Important as stone tools were to our ancient ancestors, they actually tell us about something more important than just the ability to manipulate stone and hence get more food. Stone tools represent cultural activity. Although some aspects of our understanding of Paleolithic cultures comes from anthropological observation of forager societies that have survived to modern times, and are therefore somewhat hypothetical, the tools tell us certain things must be true. To start, stone tools tell us that our ancestors worked and lived in groups. The size of the groups is more difficult to determine, but kinship and families provide a basic group that seems common to humans and many other primates, so it seems reasonable to assume that other hominins also had such groups. It also seems likely that the skills associated with stone tool making had to be spread among the members of such groups. If only one member could make stone tools, the group would be in great danger if that individual died or was unable to make tools. This does not mean that there weren't members who did the bulk of stone tool work, but skills had to be spread out for convenience and safety. That would mean, in turn, that such skills had to be taught by experienced tool makers to others, and passed down from generation to generation. Thus, education was part of our ancestors' experience, since learning to copy the tool making techniques would require observation, memory and practice.

On a psychological level, the tool makers would have to keep in mind some standard of utility, or a mental picture of what a stone tool should be like. This reveals clues about mental processes such as analysis and memory. The mental template created for stone tools was very powerful, especially in the Paleolithic era when the pattern of

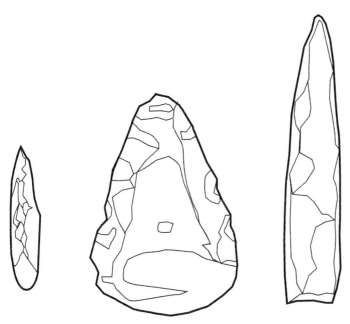

Figure 2.1 Stone tools. From left: scraper/knife, hand axe and spear tip.

stone implements remained consistent for many generations (Figure 2.1). Such mental and social skills once developed would be available for activities beyond tool making.

Put together, stone tools tell us about a society that could cooperate, communicate (even if this did not mean speech as we use it), share skills and pass them on to new generations, as well as possessing the physical and intellectual skills to manufacture stone tools in bulk and from a variety of different materials. While stone tools may have made hominin life better, it was the skills needed to make the tools that really allowed our ancestors to survive and prosper.

Fire

The next great step in tool use was the controlled use of fire. The earliest example of this may be from 1.5 million years ago. Burnt bones found at Swartkrans, north of Johannesburg, South Africa, were first described by Bob Brain and Andrew Sillen of the University of Cape Town in 1988. Samples were later tested using Electron Spin Resonance to determine the heat they were exposed to when they were burned. Since natural fires such as a brush fire burn at a lower temperature (around 300–400° Celsius) than a camp fire, the discovery that the bones were exposed to a temperature above 600° strongly suggested deliberate use of fire rather than exposure to a naturally occurring fire. The evidence, although intriguing and suggestive, is indirect. A further complication is that it is also unclear what group of hominins might have been using fire, since fossil remains of both Australopithecines and *Homo erectus* from the period have been discovered in the area.

The carbonized remains of seeds and fruit from between 790,000 and 1 million years ago give stronger evidence of deliberate use of fire, and it seems likely that *Homo erectus* could use fire. By 250,000 years ago, its use was widespread, with clear evidence from sites in Europe, Africa and China.

Like stone tools, the use of fire offered our ancestors some immediate uses and some less obvious benefits. Fire made a wider range of foods available for consumption, as well as making food, especially meat, safer to eat. In terms of nutrition, cooking food can decrease the amount of vitamins, but also vastly increase the protein available, and what was lost in specific nutrients was made up for by the increased amounts of food eaten. The biological anthropologist Richard Wrangham pointed out that eating a raw egg gained 51 percent of the protein, while a cooked egg provided 91 percent (Wrangham 2009). In addition to the extra energy, we also gained time. Our primate relatives spend a great deal of time chewing and digesting raw food, while humans have the smallest digestive system to body size of all primates. We may have evolved because cooked food is easier to digest and provides more energy.

Eating became a more complex activity as food was brought back to an existing camp or a fire was started to process food at the place it was found. Human society was formed around fire and eating. Fire also allowed food to be preserved more easily by drying and smoking, and this in turn made it more portable and longer lasting, which in turn increased the range and duration of activities such as hunting, storing food for periods of need, and migration. Fire provided warmth and light, increasing the parts of the globe that could be inhabited. It also made our ancestors safer from predators.

The less obvious benefits of fire were the lessons it taught about the transformation of materials, the consumption of fuel and the temperament to make and tend fires. The transforming power of fire runs from the simplest hardening or shaping of wood by careful heating to the magical transformation of mineral ore into metal. It is likely that over the generations every kind of material and object our ancestors had around them was tossed into a fire to see what would happen to it. In some cases the results were unimpressive, although even hot rocks could be used to boil water in a skin sack or act as a rudimentary skillet. In other cases, the results were more spectacular, such as the breaking of rock, the cooking of eggs, intoxication from the smoke of certain plants, or the firing of ceramics.

Fires had to be tended, and that required organization to keep a supply of fuel available. With the demands of keeping a fire going came a different sense of time as the night was pushed back by the artificial light. This, in turn, increased the time for social activities as people spent hours around the fire between the activities of the day and sleeping.

The consumption of fuel meant that keeping a fire going required planning and fore-thought. Fuel had to be gathered and that taught the lesson that a certain amount of material consumed meant a certain amount of time. It is likely that the banking of fires (covering the hot coals with dirt to preserve them for later use) was learned as a means to lower wood consumption and reduce the need to tend the fire throughout the night, adding another duty to the regime of fire.

Self-Portrait or Talisman

The properties of fire were likely also associated with magic and religion. There is evidence that ancient people burned plants that are known to produce scented smoke and as hallucinogens, suggesting ritual activity, and the long history of tales about the gift of fire from the gods undoubtedly extended back to prehistoric times. It is difficult to know for certain what religious activities took place among our ancient ancestors; for example, there are clay, bone or stone objects depicting animals and people, but whether the figures were simply craft items or talismans, or had deeper religious sig-nificance, is impossible to determine. In particular, a variety of female figures known as Venus figurines have been the basis for a great deal of conjecture about the worship of a goddess figure, especially one associated with fertility and birth. Although it is still a controversial claim, the earliest example may be the Venus of Berekhat Ram, dating to between 200,000 and 300,000 BCE, while the most famous, the Venus of Willendorf, dates from around 22,000–24,000 BCE. Found in 1908 near the Austrian city of Krem, the Willendorf figurine was carved from limestone and colored with ochre. The time and skill necessary to produce these figures suggest that they were important and highly valued objects, but whether they depict a stylized goddess and were objects of veneration, were talismans betokening fertility or safe childbirth, were portraits of actual people or had some other significance we will never know.

The skill necessary to create such figures does, however, tell us about a society that valued such artistic endeavors and was willing and able to devote time and effort to such activities. To date, the earliest items of personal decoration, small shells pierced and strung together to form a necklace or bracelet, come from about 75,000 BCE. They were discovered by Christopher Henshilwood and his team in the Blombos Cave in South Africa in 2004. It was also at Blombos that a small block of ochre with finely scribed lines was found, strongly suggesting that art and decorative skills played a sig-nificant role in the life of our ancient relatives. These items also tell us about the ability of the creators to plan ahead and a desire to transform matter for both practical and aesthetic reasons.

The Upper Paleolithic Revolution

Part of the reason there was increasing human activity was the ending of the last glacial period. Through geological time, there have been many periods of temperature fluctuation, with a warm period in the Pleistocene ending around 110,000 BCE, and the planet not reaching pre-glaciation temperatures again until around 15,000 years ago. There is much debate about the relationship between the great climate changes and early human history, but about 30,000 years ago there was a sudden burst of innovation in tool development and evidence of social change. Our direct ancestors have been around for some 250,000 years and during that time there were changes in the basic tool kit, but during the warming following the last glaciation there was suddenly a far greater variety of tools and other objects, *and* a greater disparity in the types of tools found among geographically separated groups of people. In comparison, all Neanderthal groups of the same era appear to have had basically the same tools regardless of where they lived. What was once a one-knife-fits-all approach started to give way to tools designed for specific environments and specialized tasks.

This change is sometimes referred to as the "Upper Paleolithic Revolution." Some of the leap forward was associated with decorative work on tools, the appearance of cave art and the making of beads and other adornments that did not seem to exist in the earlier period. There is controversy about when, where and what exactly changed, particularly with the discovery of early examples of artistic activity pushing back the beginning of artistic aspects of tool using societies, but the most contentious issue is why there was a sudden and profound change in human cultural activity in comparison to the long period of stability that preceded it. In addition to the greater diversity of tools, including the appearance of harpoons and other fishing tools, there was a marked increase in evidence of social activity such as ritual burials, the manufacturing of clothing using sewing needles, the first indication of record-keeping using tally sticks of bone or ivory, evidence of trade over long distances, and a blossoming of decorative art. Two ideas have been advanced by researchers to explain the leap. The first theory is that environmental changes forced humans to diversify their behavior. Climate changes isolated different groups and changed what they needed to do to survive. Such diversification might have remained isolated, but as the planet warmed and people moved into new regions, those activities and inventions that had developed to meet local conditions were transmitted to a wider community, leading to new modifications.

The second theory is that there was a biological trigger based on a change to the brains of *Homo sapiens*. The "new brain" theory suggests that changes in the structure and function of the brain led to the ability to conceptualize the world differently, and in particular engage in abstract thought that enhanced planning for future events and the memory of past events. Part of the debate is also based on the degree to which humans and our closer relatives used language. There has been a general assumption that the higher the language level (especially the ability to communicate abstract ideas) and the greater the use of language, the more complex the culture of the speaker.

The debate over language has also been part of the discussion about the relationship between Neanderthals and *Homo sapiens*. The two hominin groups overlapped chronologically and are closely related genetically, but humans did not come from Neanderthals. Both groups used tools and fire, and hunted in groups. Physically, Neanderthals were more robust, but they faded from the historical record, leaving *Homo sapiens* as the only surviving hominin from the *Homo* line. For many years, there was an idea that Neanderthals lacked the physical structures to produce speech and it then followed that the Neanderthals could get to a certain social and technological level and no further. New anatomical work suggests that Neanderthals did have the structures to produce speech, although there is no way to know if they did use it as we do.

It may be that both theories contain part of the answer, since there is a relation between environmental factors and genetics, and between cultural activities and survival. Natural variation may have produced genetic shifts that changed *Homo sapiens* brains, offering a better adaptation in the local environment, while at the same time the environmental changes created the conditions that allowed such genetic modification to be spread through the larger population. Although a precise answer as to why there was a Great Leap may never be possible, it is clear that our ancestors developed an increasingly technological mind, one that demonstrated a powerful ability to manipulate the material world, creating new devices such as tools, weapons and decorative items. The technological mind opened new horizons in communication, social organization and the arts.

There are various theories about what happened to the Neanderthals. Although genetic evaluation has revealed that there are traces of Neanderthal DNA in some modern people, it does not seem likely that *Homo sapiens* interbred and thus swamped out the Neanderthals. Given the generally belligerent nature of humans, it is easy to think of direct confrontations between humans and Neanderthals as at least part of the reason for their disappearance, but evidence for this is very limited. Humans did take over places that had been used by Neanderthals, but whether it was vacant occupancy or eviction is unknown. One of the only things we know for sure is that *Homo sapiens* made a greater variety of tools and thus could exploit more natural resources than Neanderthals.

⚙ Art

Although a stone scraper or a spear can tell you a great deal about the life of the person who created it, it is through the arts that we gain a sense of what our relatives were like as people and a community. By 25,000 BCE, art can be found at many sites.

Of particular interest for both their artistic and technological resources are the various cave painting sites. More than 350 have been discovered worldwide, with the most spectacular found in France and Spain. Although there are ongoing arguments about the dating of the pictures, the oldest site discovered is probably Chauvet Cave in southern France. The dates for the pigments used in the cave art run from about 29,000 BCE to as old as 32,000 BCE. Discovered in 1994, the site differs from others in that it contains pictures of a wider variety of animals, including predators, whereas at other sites the artists focused on the animals that were the target of the hunt. It is clear that a great deal of work was involved in creating the images. In addition to the actual time it took to paint the hundreds of figures and designs, there was also time spent collecting and processing the pigments, scraping and cleaning the walls, and setting up the fires to illuminate the work. There is also evidence that a number of the images were repainted, in some cases many times.

Pictures by their very nature tell us about the ability to think in the abstract, since the image calls to mind something that is not present. While figurative work may show us the animals and hunters known to the artist, the decorative work tells us about people who were concerned with changing the appearance of something to fulfill some need, such as worship, memory and teaching, identification, or the invocation of magical powers. Or simply to make the decorated object more attractive.

Cave painting also suggests that dance and music were part of Stone Age life. A cultural treasure-trove was uncovered in 2008 at the Hohle Fels Cave in southern Germany, near the city of Ulm. This discovery has given a new insight into the social life of Upper Paleolithic humans. Three delicate ivory flutes were found, along with flint tools, and mammoth, horse, reindeer and bear bones. A female figurine was found in the same layer, as were the constituents of paints, made from various minerals, charcoal, blood and animal fats.

The flutes show complex manufacturing technique, where the maker first cut the shape from a solid piece of tusk, then split it lengthwise and hollowed it out. Holes were drilled, and the two halves glued back together. A pair of V-shaped notches at the top created an instrument much like a penny whistle. Marks on the flutes suggest that the placement of the finger holes may have been determined by a template, or by copying an existing instrument.

Nicholas Conard, one of the team that analyzed the discoveries, pointed out that the significance of the flutes goes beyond the skill of their manufacture, and could be evidence of large social networks (Conard, Mainat and Münzel 2009). Musical instruments suggest a process of education, group activities such as playing in groups, telling stories with music, singing and dancing. The instruments may have played a role in ceremonies and celebrations. Such activities create the cultural foundation for society. In addition to the skills needed to make such items, it tells us of the values and interests of our ancestors. While music by itself did not directly lead to the spread of humans around the world, the intellectual capacity to make and play such instruments demonstrates the capacity to work together to overcome other challenges.

Ötzi and Evidence about Neolithic Life

The sewing needle was also important. Delicately carved from bone, needles tell us about the life of Stone Age societies, suggesting serious effort was expended on domestic technology. Not only were clothes important, but tailoring for fit and decorating them were part of culture. The discovery in 1991 in the Alps of a mummified Ice Man, often known as Ötzi and dated 3400 and 3100 BCE, as well as finds of clothing fragments from a number of other sites, revealed that sewing skills were highly refined by the end of the Neolithic. Although the fragments of such perishable items as clothing are rare, if Neolithic cultures shared similar interests with later cultures, decoration represented artistic ideas, conveyed status and had symbolic meaning.

Along with tailored clothing, from the Paleolithic onwards, there were other forms of adornment such as beads used for necklaces and bracelets, and pendants. These were made of a large range of materials, from stone and bone to shell. One of the earliest such objects was an amulet found in Hungary carved from a mammoth tooth and dated to about 100,000 BCE, making it likely the product of a Neanderthal artist. By 30,000 BCE beads were a common object found in early human settlements. They also provide evidence for later trade since decorative materials have been found long distances from their likely source of raw materials.

Many of the cave paintings show scenes of hunting, and the tools of the hunt were changing in this period. Spears had been used for generations, but around 13,000 BCE comes the first evidence of a boomerang. Found in Poland, this boomerang was made of mammoth tusk. Hunting boomerangs or hunting sticks have been found in Africa, Europe, Asia, Australia and North America. Although they are bent sticks (or, in the case of the Polish boomerang, made from tusk) with a curved upper surface and a flat underside, they differ from the stereotypical boomerang in that they do not return to the thrower. They are designed to fly a long distance close to the ground to stun or kill small game.

The transition from the use of a pointed stick to a spear cannot really be dated, but the spears in use by humans indicate that hunting larger game was part of early food gathering. Hunters with spears shown in cave paintings indicate that hunting large game was not a solitary activity, but required coordinated effort. The logistics of tracking, stocking and killing large animals was also time-consuming, and dangerous. Hunters often came back empty-handed, but when they were successful, the hunters gained far more than just food. Large animals became an important source of materials, with every part of the animal being used. Muscle and organs were food, while bones, horn and teeth were utilized in a wide variety of tools and decorative objects. Intestines and sinews made cord and binders. Hide could be used for a huge range of items such as clothing, rope, footwear, containers and building material.

Hunting

In addition to taking on individual large game animals, hunters developed methods to hunt groups of animals using wallows, jumps and corrals. Herd animals such as bison would be lured or driven into these traps. Wallows were shallow water such as ponds, the edges of lakes and rivers where animals would get mired in mud and exhaust themselves. In northern regions, ice-covered lakes or ponds made excellent wallows, since dead animals would freeze, preserving them until they could be used. Jumps were small cliffs that were high enough to injure the animals, but not kill them, while corrals were, as the name suggests, fenced-in areas that could be closed once the animals were driven inside. This type of hunting, especially the corral version, may have contributed to the domestication of animals as the hunters kept some animals alive for later use.

Such traps could be used by small groups of hunters, but also allowed for mass hunting. The idea of mass hunting was to collect a great deal of material all at once, and the process required numerous hunters and people to process the animals. Such activities probably brought together related but independent groups for the hunt. One of the great benefits of wallows, jumps and corrals was that animals were kept alive until the hunters were ready to process them. One of the most famous jumps is Head-Smashed-In Buffalo Jump in southern Alberta, Canada, in use over many generations since about 4000 BCE and now a World Heritage Site. There is evidence of industrial-scale use, when hundreds of animals at a time were hunted and butchered.

For hunting in the field, the spear was modified by the addition of the atlatl or spear throwing stick. Examples of atlatls date back to 30,000 BCE. A logical extension to the thrown spear or dart, the atlatl was a stick or board about 30–100 cm (1–3 ft) in length with a grip at one end and a hook or notch at the other end, and worked by in effect extending the arm and adding more leverage to the hunter's throw. Although hitting game using an atlatl would have taken practice, the advantage of the greater velocity meant greater penetrating power and distance. The atlatl has been found on all the habitable continents, and was used as late as the Bronze Age in battle. It continued to be used by the indigenous people of Australia and parts of Africa at the time of European contact.

Another form of spear, the harpoon, was also developed sometime between 30,000 and 20,000 BCE. Spears were undoubtedly used to fish earlier than this, but the harpoon modified the point and the shaft to secure the fish. Harpoons indicated a change in the hunting of water creatures as people began to pursue whales and other large animals, not just fish in streams and at the shore. Like the hunting of large land animals, large aquatic animals provided a wealth of materials, from fur, hides, bones and teeth from otters and walruses, to baleen from whales, and teeth and hide from sharks. These hunts also required boats and highly organized groups of hunters.

The ultimate Stone Age hunting tool was the bow. The origins of the bow and arrow are hard to determine because no clear evidence of bow construction has survived to the present, although the existence of bows can be inferred from what appear to be stone-tipped arrows in Parpallo, Spain, dating back to around 20,000 BCE. Stone arrowheads became one of the most common Neolithic artifacts, with tens of thousands of stone points found around the world. There is an interesting question about the utility of stone points over simply using sharpened wood arrows. Modern tests have suggested that arrows with stone points were not significantly more deadly or accurate than wood points, so it would seem like a waste of time to make stone arrowheads. This suggests that stone points may represent an early case of technophilia, or the desire to have the latest technology whether it was really better than what existed. It is also likely that while a stone arrowhead may not penetrate much more deeply than a wood point, the shape makes it harder to dislodge the arrow after it hits, and it would continue to cut tissue within the wound after the initial impact. Since it was unlikely that a single arrow hit on an animal would kill it, the progressive damage of stone-tipped arrows may have justified the additional time required to manufacture them compared to sharpened wood.

Although arrowheads exist in large quantities, the earliest evidence of an actual bow comes from around 8000 BCE. By 5000 BCE the Egyptians were proficient with the bow. It seems to have replaced the atlatl in most parts of the world as both a hunting tool and a weapon of war at about the same time. It offered greater accuracy and a higher rate of fire, as well as advantages in the manufacturing of the arrows over the larger spear or dart. The bow was also better suited to forested and other conditions where finding space to throw a spear would be problematic. The bow would be an important hunting tool and weapon of war until it was replaced by gunpowder weapons starting in the fifteenth century.

✿ Pottery

Returning to the period when the cave painters were at work, there was the appearance of the first use of pottery. The earliest known ceramic objects were discovered at Dolní Věstonice in the Czech Republic. The Venus of Dolní Věstonice figurine has been dated to 29,000–25,000 BCE. It is likely that clay was used much earlier than this, but it was not fired and examples have not survived to the modern day. Although the early surviving ceramic objects were heated in a fire, the use of pottery for durable items such as pots and bowls did not happen until about 10,000 BCE. The technical advance that made ceramics possible depended on a greater ability to control fire, using a pit kiln rather than an open fire, and thereby raising the temperature from around 500° Celsius to over 1000°.

The next great transformation of human society occurred between 14,000 and 10,000 BCE. That was the beginning of agriculture. Like so much of our early history, it is not certain exactly where agriculture got its start, but the earliest evidence of deliberate and sustained plant growing was in the regions of the Fertile Crescent and Levant, running from the Mediterranean around modern-day Turkey to Egypt in the west and along the Tigris and Euphrates rivers to the Persian Gulf in the east. Even in this region, there is evidence that agriculture was introduced multiple times, so it was not a single discovery at a single time. The discovery of plant cultivation meant not only a new set of tools, but a profound change in psychology.

1 What does the long-term production and consistent design of stone tools tell us about the people who made the tools?
2 How did the controlled use of fire contribute to the development of new forms of technology?
3 In what ways did climate contribute to the development of new tools by our ancient ancestors?

FURTHER READING

New discoveries in anthropology and human paleontology are made with such frequency that any examination of ancient technology must come with a warning label: Subject to change without notice. A very accessible introduction to the subject of our earliest ancestors is Carl Zimmer's *Smithsonian Intimate Guide to Human Origins* (2005). One of the most significant areas of research in the history of technology concerns the reasons why new tools were created. Ofer Bar-Yosef, Harvard professor and curator of the Palaeolithic Archaeology collection of the Peabody Museum of Archaeology and Ethnology (www.peabody.harvard .edu), discusses one the earliest periods of tool invention in his article "The Upper Paleolithic Revolution" (2002). The controlled use of fire was a major turning point in human history, but Richard W. Wrangham argues in his book *Catching Fire* (2009) that it was even more important than previously thought because it changed human evolution. The discovery of Ötzi, a man frozen and preserved around 5,000 years ago, is examined by David Murdock and Bonnie Brennan in the PBS *Nova* documentary *Iceman Reborn* (2016).

3 Origins of Civilizations

16,000 BCE	Dogs domesticated
11,000–9,000	Sheep and goats domesticated
10,000	Grain cultivation
9000	Göbekli Tepe site
7500	Çatalhöyük settlement
7000	Mass production of pottery
5000	Horse domesticated
4241	Calendar in Egypt
4000	Bronze
3150	Unified Egypt
3000	Cuneiform writing
2560	Great Pyramid at Giza

The transition from hunter-gathering societies to settlements required new tools and new ideas. The key to the new way of life was agriculture, but the discovery of agriculture also required a change in human psychology, so it is not surprising that there was a transition period when settlements appeared and disappeared. When agriculture was understood well enough to make it possible to support permanent settlements, there was a growth in populations. Larger populations required more people to manage the society, but food surpluses allowed people to have specialized roles such as potters, scribes or soldiers. Job specialization led to the freedom to invent new tools, and larger settlements meant new kinds of problems such as the storage of grain or the construction of temples. New tools also gave our ancestors a greater ability to control the environment. The first empires emerged on large river systems that provided good agricultural land and transportation. With larger populations and job specialization, the river empires began to undertake large projects such as building temples or pyramids. These kinds of mega-projects and the scale of the empires led to a growth in bureaucracies to control and tax the population. Bureaucracies required mathematics and writing to keep track of all the complex details of life in large settlements and larger empires.

There are various theories about the transition from foraging societies to agricultural societies. To understand the transition, some aspects of foraging are important to note. First, although the hunting aspect was important for food and materials such as bone, hides and sinew, it was primarily the gathering that kept the society fed. Berries, roots, tubers, edible plants and seeds as well as insects and small game were the basis of the daily diet. Fishing and hunting for medium to large game was often seasonal or simply not reliable enough for daily consumption. Foragers, particularly if they were nomadic or semi-nomadic, could only store and carry a limited amount of food, so the basic rule was "eat it when you get it." It is likely that the gatherers made the observation that plants tended to grow in specific localities and that there was some relationship between seeds and mature plants. There may also have been the observation that useful plants tended to grow up around camp sites as waste or inedible pits and seeds along with undigested and excreted seeds were discarded nearby. At some point (and probably many times), people recognized that a seed led to a plant that produced seeds and so on. This observation was part of Robert Braidwood's "Hilly Flanks" hypothesis that intensive gathering of wild grains found around the Taurus and Zagros mountains (running from western Iran through Iraq to the southern part of the Persian Gulf) led to the discovery of the process of planting and cultivation (Map 3.1). Other models depend on growing demands for food, either because of the demographics of population growth as proposed by Carl Sauer and others, or because of feasting as suggested by Bryan Hayden (Sauer 1952; Hayden 1992).

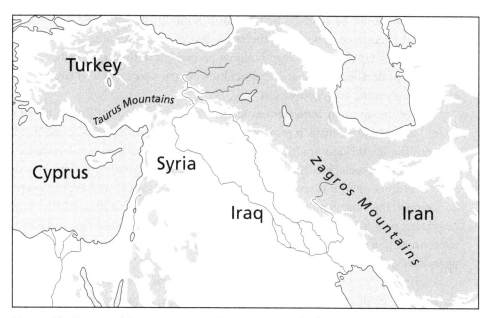

Map 3.1 The Taurus and Zagros mountains. These regions may be the original sites of agriculture.

Until recently, it had been assumed that agriculture was the impetus to the crea-tion of permanent settlements. This theory, although logical, has been challenged by the discovery of pre-agrarian settlements. An intriguing archeological site that may cast new light on the relationship between food gathering, religion and the creation of settlements is Göbekli Tepe, in the mountains of southeastern Turkey. The earliest layer dates back to about 9000 BCE, making it one of the oldest prehistoric construc-tion sites. Although only a tiny portion of the site has been excavated, the discovery of many carved pillars suggests that it was an important cultural site, perhaps indicating religious worship. One of the most interesting aspects of Göbekli Tepe is that, although there are piles of bones from food animals, no residential areas have yet been uncov-ered at the earliest level. Klaus Schmidt, who began the main excavation of the site, believes that Göbekli Tepe was a sanctuary with religious importance and in turn led to the creation of settlements. Schmidt said of the site, "First came the temple, then the city" (Schmidt 2000). The idea that cooperation and construction for ritual reasons preceded practical utility reversed the long-held idea that more structured religious practice followed the emergence of agriculture. The discovery of many other ritual sites in Turkey, Jordan and Israel that seem to predate the establishment of domestic struc-tures supports this theory. The demands for food created by the concentration of peo-ple at the site may have contributed to the creation of agriculture as people cooperated to guard and harvest wild grain which grew in the region. According to this model, agriculture was not created incrementally, but as a response to a specific demand for food that could not be supplied by traditional hunting and gathering. In addition to learning the basics of agriculture, the skills developed by building the structures at Göbekli Tepe were then used when building permanent settlements. At the very least, Göbekli Tepe demonstrates that pre-agriculture people were capable of large-scale and long-term planning and cooperation.

Another pre-agrarian settlement is Çatalhöyük, discovered in Turkey in 1958 by James Mellaart and his team. It provided stunning evidence of a sophisticated Neolithic culture.[1] The site includes many buildings, examples of ritual burial and evidence of a rich cultural and social life. The earliest settlement layer at Çatalhöyük dates to around 7500 BCE, and even if it is not the oldest permanent settlement we have discovered, it does remain the most populated site of its age, with an estimated 10,000 inhabitants at its greatest population point. It was a collection of mud-brick buildings, and over time the construction became so dense that many of these could only be entered by climbing up ladders and using hatches in the roofs.

In 2017, Ian Hodder and his team completed twenty-five years of excavation at Çatalhöyük. Hodder believes that they may not have uncovered the earliest structures at the site. Çatalhöyük raises many questions for archeologists and historians. Why did it come into being and why did it eventually fail? What led to such a high concentration of population, far larger than any other site discovered so far? Why did the people at Çatalhöyük build such a dense settlement rather than expanding around the perimeter? As time went on, newer structures were built on the site as the settlement went through periods of decline and resettlement. The culture developed a more agrarian model, but despite the large population of Çatalhöyük, it remained a Neolithic culture mostly lacking the crucial technological innovation that would separate the Neolithic agrarian settlement from the early urban settlement: task specialization. Although a few tasks at Çatalhöyük appear to have been undertaken by artisans, most tasks were done by households. Everything from construction, plastering, tool production and food gathering were carried out in each home, rather than being done by specialists such as craftspeople and artisans in purpose-built buildings.

The success and the failure of Çatalhöyük may have been the product of the same thing, namely the power of a settlement to efficiently exploit resources. This allowed for a far greater concentration of people in one place, but such a large population also put a strain on the resources of the area. A lack of wood for fuel and construction, along with evidence of malnutrition and high infant mortality, suggests that the people of Çatalhöyük lived a subsistence life, right at the edge of what could be supported by the resources of the region. Any fall in productivity, as likely happened when there was a global cold period around 6200 BCE, was catastrophic. As family or kinship units struggled to get enough food, fuel and other resources, they were in competition with each other for a static or shrinking resource base. Although the story of Çatalhöyük has not been completely deciphered, it seems likely that some people could have moved away when problems first developed, but the majority remained until forced to leave. Those who stayed were held by bonds of tradition and family ties, the general stability and protection offered by the settlement even as life became more difficult, and the limited number of alternate sites to establish a new settlement.

Agriculture, in a more general way, was a response to need, but Hodder's work at Çatalhöyük suggests that it took thousands of years to shift from hunter-gathering to an agrarian life. As demand for food grew, new strategies for producing more food from the same area of land were necessary and so cultivation replaced gathering. Once the process of cultivation was established, people begin to selectively breed plants to enhance desirable characteristics such as size of fruit or quantity of seeds. To do this required not just a change in technique from hunting and gathering to cultivation; it required a shift in psychology as successful farmers had to give up the principle of "eat it when you get it" and institute the rule "save the best for later." Farmers who ate all the biggest seeds and then planted the small ones might literally breed themselves into starvation. This also prompted efforts to preserve food, so the process of storing, cooking, air-drying and smoking expanded and extended the food supply.

Ultimately, salting food offered the best method for preserving food, and finding and controlling sources of salt shaped civilization. Almost every major settlement in ancient times had access to salt either from mines or from a process of brine evaporation. The settlement of Solnitsata in Bulgaria was built around a salt mine and flourished about 4500 BCE, while the Incas operated evaporation ponds at Maras, Peru.

The climate played an important part in the creation of agriculture as well. As the globe warmed, wild cereals became more abundant. They were annuals that produced seeds that could survive dry seasons to propagate when growing conditions were right. In the regions where they grew in large quantities, they offered an easy and convenient source of food. Two types of wheat, emmer and einkorn, were early cultivars from the area around Karaca Dağ in southeastern Turkey (near the site of Göbekli Tepe). Then came barley, peas, lentils, bitter vetch (a legume), chick peas and flax. The large-scale production of cereals required the invention of an ever-increasing range of goods and systems such as storage, cooking techniques and implements, and farming tools.

In many places, cultivation of grains was done in tandem with hunting, but in the most fertile regions the advantages of agriculture resulted in such a convenient way to produce food that hunting became a supplemental activity, and eventually an activity undertaken for special occasions, or for reasons of class, religion or social standing rather than as a necessary source of food. In many places, hunting for large game would be restricted to the upper class and taking prohibited animals became the crime of poaching.

The basis of grain agriculture was simple: gather seeds at the end of the growing season, save some of them until the start of the next growing season and scatter them on the ground. Wait, try to keep animals from eating too much of the crop while it is growing, hope the weather is good and in the harvest season gather the grain. Then separate the kernels from the hulls, store and repeat. In some regions of the world, such as Cambodia, there could be two or even three harvests a year, but for the most part agriculture was an annual event, following a spring/fall or wet/dry season planting and harvesting pattern.

Table 3.1 Domestication of plants.

Date BCE (approximate)	Name	Likely region
10,000–9000	Emmer wheat	Turkey and Fertile Crescenta
	Einkorn wheat	Turkey and Fertile Crescent
	Barley	Fertile Crescent, Asia
8000	Potatoes	Peru
	Pumpkin	North America
	Beans	Peru, South America
	Rice	Indochina
7000	Durum wheat	Turkey
	Sugar cane	New Guinea
	Yams	Indonesia
	Bananas	Indonesia
	Coconuts	Indonesia
	Flax	Fertile Crescent
	Maize	Mexico
	Peppers	Mexico
	Lentils	Fertile Crescent
6000	Citrus fruit	Asia
	Millet	North Africa
	Peach	China
	Avocado	Mexico
5000	Date palm	India
	Cotton	Mexico
	Grapes	Turkestan
4000	Sorghum	Sudan
	Olives	Crete

At about the same time that plant farming was starting, there was another amazing development that helped humans conquer the world. This was the domestication of animals (Table 3.2). Although it may seem strange to think of animals as technological objects, they represent an important solution to several real-world problems and demonstrate the ability of people to learn about the world around them and use that knowledge to exploit resources. Entire cultures grew up around the domestication and use of animals, and domestic animals are a major factor in the world economy to this day.

Animals are a source of food and other materials such as hides and bone, making them valuable objects for our ancestors, but in fact the first animals to be domesticated were dogs, not food animals. There is almost no aspect of the issue of canine domestication that has not been hotly debated, such as where dogs were first domesticated and brought under human control, what animal was the precursor to the modern dog, and why dogs could be domesticated. Further, even though humans have a great deal of control over dogs, the relationship between humans and *Canis familiaris* is different than with most other domestic animals, representing something closer to a partnership, with dogs allowed a degree of autonomy and familial contact unlike any other domesticated animal, with the possible exception of the domestic cat.

Table 3.2 Domestication of animals.

Date BCE (approximate)	Name	Likely region
16,000–14,000	Dog	China, perhaps Mediterranean
11,000–9000	Sheep	Fertile Crescent
	Goat	Fertile Crescent
8000	Cat	Fertile Crescent
7000	Pig	Turkey
	Chicken	Southeast Asia
6000	Cow	Fertile Crescent
5000–4000	Alpaca, llama	Andes
	Horse	Ukraine
4000	Camel	Asia
	Donkey, mule	Egypt
	Water buffalo	China
3000	Reindeer	Russia
	Elephant	India

The use of modern DNA analysis is offering some new insight into the course of human–canine relations. Although there are remains that indicate very early domestication from about 33,000 years ago, when dogs and wolves diverged genetically, the best evidence suggests that the modern dog is most closely related to Asian wolves, and that domestication can clearly be seen in Asia, between 16,000 and 14,000 years ago (Figure 3.1). Although all dogs are genetically similar, not all dogs are from a single genetic ancestor, suggesting that domestication was not a single event.

Although the Asian model seems the strongest from a DNA point of view, it is not conclusively proven, and there is archeological evidence for placing domestication in the Mediterranean basin. The discovery of burial sites from 12,000 years ago in Israel that had human and dog bones together suggests that dogs were an established part of the community. Conflicting evidence is frequent and may be the result of multiple cases of domestication that happened in different places. For example, the dingo, or wild dog of Australia, may have arrived in Australia from Southeast Asia as early as 18,000 years ago or as late as 4,500 years ago, and could only have arrived there with human help, raising the question of how and why it returned to a feral state. What is more certain is that by 9000 BCE dogs were part of human life, and they went wherever humans went. Dogs were used to help hunt, to guard people and property, and to guard and herd other domesticated animals. They were also used to pull travois and snow sleds and occasionally eaten as food.

The basic process to attempt to tame animals was to capture a specimen, particularly a young one, and tie it up or confine it. Feed the animal and hope that over time it accepts the food and the source of the food. Some animals could be domesticated and would eventually be so selectively bred that they could not return to the wild. Ruminants such as goats, and other herd animals proved relatively easy to domesticate. Some species could be domesticated to varying degrees, such as domestic cats

Figure 3.1 The domestic dog. The closest relative in the wild is the Asian wolf, but generations of selective breeding have created a symbiotic relationship between humans and dogs.

(perhaps as early as 9000 BCE and significant in Egypt by 4000 BCE), and hawks (starting around 2000 BCE). Others could not be domesticated, regardless of effort, such as jackals or hippopotamuses. The hyenas, which the Egyptians attempted to domesticate around 2500 BCE, would never accept human control.

After the dog, the next animals to be domesticated were goats and sheep (which are very closely related and therefore difficult to distinguish in the archaeological record). The earliest evidence of domestication comes from the Fertile Crescent some time before 8000 BCE. Water buffalo, pigs and cattle were added to the stable between 6000 and 7000 BCE. With the addition of each animal, humans gained resources and had to learn new skills. Animal husbandry, or the care and management of farm animals, became a major occupation. In turn, husbandry led to the concept of selective breeding, whereby humans chose the characteristics they wanted to enhance or suppress and by controlling reproduction tried to shape animals.

The first agricultural revolution, created by the discovery of planting and the domestication of animals, led to the greatest cultural revolution in human history. This was the origin of civic society, or a society based on permanent dwellings established where agriculture had become the central means of food production. As with agriculture, the creation of settlements appeared in a number of locations and over the generations some settlements faded away, some were destroyed and others survived. In many cases, settlements were built on the location of previous ones, leaving us a kind of time machine as layer upon layer of human habitation can be peeled back to reveal the past. Although there continues to be debate about the origins of permanent settlement, there are several common factors that were necessary to establish such settlements. The first aspect was geographic. Settlements had to be located on or near water and where the food would grow. This meant that the shores of lakes and rivers, and especially delta regions, were prime locations. Other considerations also played a role in the placement of settlements, such as the availability of salt, access to sources of wood, stone or other building materials for construction, and access to fuel for heating, cooking and crafting. Supplies of wood for construction and fuel were useful, but some settlements had to import wood, often from long distances, either because local sources were exhausted or because trees did not grow naturally in the area. Other factors such as mineral deposits, geological features and religious sites also influenced the location of settlements.

It is likely that the very earliest sites went through a transitional period that saw camp sites made more permanent by modification of the terrain such as leveling and adding drainage, leaving tent posts or frames for shelters, constructing fire pits, and gathering brush to create fences. Later, some sites also had walls of wood or stone. In other places, caves were modified to increase size and accessibility.

Although the consistent production of a food surplus does not guarantee the success of a settlement, no significant civilization could exist without it. The people of Çatalhöyük ended up in a kind of technological trap. Agriculture allowed for stability and population growth, but as the population grew, the potential for surplus decreased. Without a surplus, there was no way to free community members from the daily activities of subsistence living to undertake specialized tasks. Without specialization it was harder to use resources efficiently or to plan ahead. Without the safety net of a surplus, innovation was difficult or impossible to conceive or implement. Since it does not appear that Çatalhöyük failed because of a sudden catastrophe such as an earthquake or invasion, it seems likely that the settlement faded from history, literally and figuratively having eaten itself out of house and home.

Another of the earliest habitation sites has been found at Tell es-Sultan, near Jericho, on the shore of the Dead Sea. The earliest remains are the foundations of stone structures dating back to around 9000 BCE. By 8350 BCE there was a stone wall, a tower and mud-brick buildings at the site. By that time the settlement was making the transition from hunter-gathering to agriculture and had an estimated population between 2,000 and 3,000. They grew emmer wheat, pulses and barley. The construction of Tell es-Sultan suggests a much higher degree of job specialization and community work than seen in Çatalhöyük and also indicates that the settlement was producing surplus food, an important factor for craft development and innovation.

An agrarian settlement that can consistently produce about 2 percent more food than it needs for consumption and the next season's planting can protect itself against temporary problems, but has little buffer against prolonged problems or to support many non-agrarian members of the society. When a society can produce 5–6 percent more than it needs, it can start to free people from agricultural labor. These people became the skilled craftspeople, artists, soldiers, priests and administrators. With their appearance came the transition from Neolithic agrarian settlements to towns and cities.

One of the most significant developments that indicated the transition was the appearance of fired and mass-produced pottery. Although there are ceramic objects dating back to 25,000 BCE and pots made in Japan have been dated to 10,500 BCE, the introduction of mass production of pottery largely began around 7000 BCE, coincident with the appearance of agricultural settlements. The spread of pottery was such an important technological advance that it is used as an archeological and technological dividing line. Kathleen Mary Kenyon, pioneering archeologist, introduced the terms Pre-Pottery Neolithic A (PPNA) for the period around 9000 BCE (the earliest part of Göbekli Tepe) and Pre-Pottery Neolithic B (PPNB) from 7500 to 6000 BCE (the first part of Çatalhöyük's history) to delineate the cultural and technological conditions of the early settlements. PPNA was more transitional, with grain production being supplemented by hunter-gatherer food gathering, while the PPNB period could include domesticated animals and a decline in hunter-gathering.[2] New tools appeared, but social organization continued to be based on subsistence agriculture where production was managed by family units or tribal groups.

The actual origin of fired clay pottery is not clear. Clay objects had undoubtedly been made for generations, but the transformation of clay into pottery required the conjunction of a number of discoveries that came from a combination of serendipity and the long-standing practice of putting things into fire. It is likely that the firing of clay was the effect of deliberately controlling the observed phenomenon of the vitrification of underlying clay soil when a fire was kept going for several days in a single spot. When it was discovered that heating clay to high temperatures transformed it from crumbling mass to rigid and long-lasting objects, it became important to develop methods to achieve consistent high temperatures (above 800°C and better at 1,000° or hotter). Although such high temperatures were possible with bonfires, they were inconsistent and used a great deal of fuel. Kilns, starting with pits or trenches where the pottery was placed in the bottom and fuel stacked around and over it, offered greater control and more consistent temperatures.

Suitable clays had to be identified and methods for working the clay to prepare it had to be understood. Techniques for forming the clay efficiently into the required forms had to be worked out. Slab forming, coiling, beating out and eventually wheel forming allowed the potter to create everything: tiles, platters, pots and large storage jars.

The longevity of pottery and its central place in the creation of civilization make it one of the great records for archeology. The appearance of pottery, while not as significant as mastering fire or the discovery of agriculture in the repertoire of tools that made civilization possible, was an integral part of the transition from Neolithic settlement to agrarian village. It represented the second great artisanal achievement after stone tools. It was simple, durable, and made with widely available material found along the banks of almost every river. Although the basic principles of pottery were (and continue to be) easy enough for a child to comprehend, it quickly became a skilled craft requiring years to master. Potters were the first specialized craftspeople in a mass industry and they created millions of containers, ranging from tiny ointment jars to giant grain containers. Pottery also facilitated an artistic explosion, as beads for jewelry, decorative tiles and sculptural objects could be mass-produced. Inscribed and painted decoration go back to the original invention of pottery and has provided archeologists with important insights into the technology, culture and artistic development of early civilizations.

The introduction of the potter's wheel sped up the production process enormously. Although it was once thought that the potter's wheel was somehow linked to the introduction of the cart wheel, it now seems more likely that it evolved independently as a way to make coil forming easier. In coil forming, the clay is first rolled out like a long piece of rope and then, as the name says, coiled into the desired shape and then smoothed. This has the advantage of easy construction of round objects with uniform thickness. Since it was easier to rotate the pottery than to walk around it, a series of rotatable mats, stands and eventually a platform driven by foot power were developed. The oldest identifiable remains of a potter's wheel dates from about 3000 BCE and comes from the Mesopotamian city of Ur (in modern-day Iraq), although pots that show signs of wheel forming come from much earlier. Claims for the origin of the potter's wheel cite Egypt, Mesopotamia, China, Korea and Japan as the original home of the invention. Although it may in the future be established where the earliest true potter's wheel was developed, it is likely that this is a case of co-development as artisans in many places reached similar technological conclusions to solve a common problem.

The history of early pottery is in a real technological sense the history of early civilization. Even the buildings were affected by the potter's craft as dried mud-bricks were replaced by fired clay bricks for construction. Clay roofing tiles, pipes and floor tiles would be added to architecture. Every aspect of the pottery industry had unexpected consequences: the skill with kilns would be applied to smelting metals and making glass; maker's marks, tally records and symbols to indicate contents of jars were part of the development of literacy and mathematics. Clay became the medium for record keeping, such as Babylonian cuneiform clay tablets. Storage in pottery containers preserved more food, so surpluses went up. Pottery facilitated trade in both the potter's wares and the goods that could be transported in them.

The development of pottery was closely associated with the creation of the first sedentary agricultural settlements, as the food surplus allowed people to exploit the power of biology and at the same time gave them the time to create the new tools that agriculture demanded. As pottery is heavy and breakable, it is a less suitable material for nomadic populations. Yet it was not the tools alone that allowed the successful communities to survive and grow, but a change in psychology. Communities that made the transition from the "rules" of the forager society to the new methods of agriculture also transformed their relationship with the environment, with the people of the settlement and with their tools. After learning the lesson of forward planning for the future, the next most significant psychological transformation was the sense of time and space. People whose sense of time was based on the cycle of the natural produce of the seasons, such as berry time and the movement of herds and bird migration, became tied to the seasons of planting and harvesting. Neolithic people observed the sun, moon and stars and even used such knowledge in the construction of monumental structures such as Stonehenge, but the ability to make and record long-term observations, that would in time continue from generation to generation, really developed when people stayed in one place. They developed calendars, codifying the concepts of past, present and future. By 4241 BCE, the Egyptians were using a 365-day calendar divided into twelve months of thirty days and five festival days.

Time and planning went together. Agriculture is a practice based on understanding the correct time to do certain jobs such as planting and harvesting, but those periods were broadly defined by environmental cues such as the flooding of rivers like the Nile in the spring and the ripening of the crops, not just a counting of the days. Keeping track of religious observance such as festival days, on the other hand, needed far more precision. After all, failing to worship the gods at the proper time could bring down their wrath upon the people. The skills of timekeeping could then be applied to a wide range of other activities such as schedules for building, tracking the length of contracts, recording the important events that occurred such as war, famine, and the death and birth of important people, or prediction of future events.

The concept of place acted to give humans a sense of themselves as separate from the world outside the walls (real and metaphorical) of the settlement. Place, like time, led to the creation of new skills. The idea of ownership of land pushed people to learn the skills of measurement and surveying. As settlements grew beyond the ability of the community to remember details about land ownership and use, records had to supplement memory. Disputes over land required some system of adjudication, leading to systems of law on the domestic level, and to the creation of military forces when different groups laid claim to the same land.

✿ Agriculture and War

It has often been suggested that agriculture and permanent settlements led to the start of warfare. Archeological evidence suggests that people came into violent conflict long before agriculture appeared. Raiding, capturing slaves and attempting to drive opponents from areas of natural resources were part of pre-settlement life for many of our ancestors. What settlements changed was the scope of conflict and the specialization of a small group of people as military forces, rather than hunter-warriors. What might have been a skirmish in a local conflict in pre-settlement times would become something we would see as an actual war, as agriculture provided the support for larger military forces and richer targets for attack. Further, capturing territory or defending it from capture changed the style of war, since at the most basic level the strategy of running away meant something far more serious for farmers than for hunter-gatherers. Barricades and walls that had kept livestock in and predators out became defensive architecture designed to prevent attack. While war became more serious after settlements became more prevalent, raiding, capturing slaves and attempting to drive opponents from areas of valuable natural resources continued to be reasons for conflict into modern times.

✿ Sacred Places

Although it is almost certain that there were sacred places for pre-settlement humans, settlements created purpose-built sacred places. Churches, temples, henges or altars appear in almost all early settlements. A huge amount of time and resources was devoted to the building and maintenance of these sacred structures, which in turn increased the demand for people with specialized skills that precluded them from agricultural work. Religion as we recognize it was also a product of the technology of agriculture and settlement. The requirements to appease the gods grew in complexity. Images and icons were necessary to depict and remember the various deities and serve as a focus for worship. Oral traditions were transformed into physical objects and the rites and ceremonies had to be performed by people specifically trained to enact the rituals. The priests lived in or near the temples, and over time those houses and temples became the experimental laboratory for architecture, art, record keeping, astronomy and other specialized skills.

One of the greatest examples of Late Neolithic/Early Bronze Age architecture, and home to several sacred structures, is Machu Picchu. Built around 1450 CE and located on a mountain ridge 2,430 m (7,970 ft) above sea level, it is believed to have been the estate of the Inca emperor Pachacuti. The site was abandoned around 1550, likely due to a combination of smallpox and the collapse of the Incan empire during the Spanish Conquest. The site was only known to local people until it was brought to international attention in 1911. The Inca used a variety of building methods, including adobe and fired bricks, but most of the surviving buildings were made of local stone. The construction of the walls and buildings at Machu Picchu followed classic Inca drystone construction. Stones were shaped using stone and bronze tools, and then smoothed by hammering and grinding to fit without mortar. The walls were also slightly inclined, to shed water and provide further stability. This form of construction was very earthquake resistant. In addition to earthquakes, there was frequent and heavy rain, so Machu Picchu was built with drainage systems to protect the structures from erosion.

Impressive as the architecture was, it was the organization of the population to undertake these kinds of large-scale projects that was the real secret to Inca building. Teams of skilled workers were needed to fashion and place the stones. The workers had to be fed and housed, and in the mountains that meant intensive agriculture using terraces built with the same kind of stonework.

✿ From Settlement to Empire

There are dozens of examples of early settlements such as Çatalhöyük, but four centers grew from early agricultural settlements to something greater. Agricultural surplus and predictable food production allowed for technological innovation and the creation of kingdoms and empires. We can recognize the power of these early empires in part because of their contribution to world culture, but also because they left us examples of their abilities to organize and undertake large-scale enterprises. Among the most evident aspects of regional power were the grand architectural projects such as the building of the Sialk ziggurat near Kashan, Iran, or the pyramids at Giza, Egypt.

Each of these areas of settlement was based on a river system, and from these centers came many of the roots of our modern civilizations. The great river systems were the Yangtze and Yellow rivers in China, the Indus and the Ganges in India, the Tigris and the Euphrates in the Fertile Crescent of the Middle East, and the Nile in Egypt. Each of these rivers shared a specific set of characteristics that made them ideal for the growth of agriculture in the region and civilization based on that agricultural bounty.

1 Flooding. Each of these river systems flooded on a regular basis, although the Nile was the most regular, flooding annually except in times of severe drought. This renewed the soil, so that intensive agriculture over generations could be practiced. The flooding also promoted record keeping and surveying, as farm plots needed to be set out using measurements from markers unaffected by the flooding when local landmarks were changed or washed away.
2 Navigation. The four river systems all had long sections that were navigable, allowing for transport and communication. In terms of the creation of empires, controlling the river generally meant controlling the people. This came from the obvious level of control where the rulers could determine who could sail on the river and move military forces around, but it also has a more subtle level of control as the river was the communication system of the empire. By controlling the flow of information, the rulers of the river empires could control what people knew and could do.
3 Fertile deltas and marshes. These parts of the river systems were important for agriculture, and for natural resources such as food from hunting, reeds for a variety of uses from baskets to writing materials, and clay for pottery and bricks.

The greatest and longest lasting of the river empires was the Egyptian empire, which rose along the Nile (Table 3.3). The power and long history of this empire depended

Table 3.3 Egyptian eras.

Early Kingdoms	c. 3150–2686 BCE
Old Kingdom	2686–2181
First Intermediate	2180–2061
Middle Kingdom	2061–1690
Second Intermediate	1674–1549
New Kingdom	1549–1077
Third Intermediate	1069–653
Late Period	672–332
Ptolemaic	332 BCE–30 CE

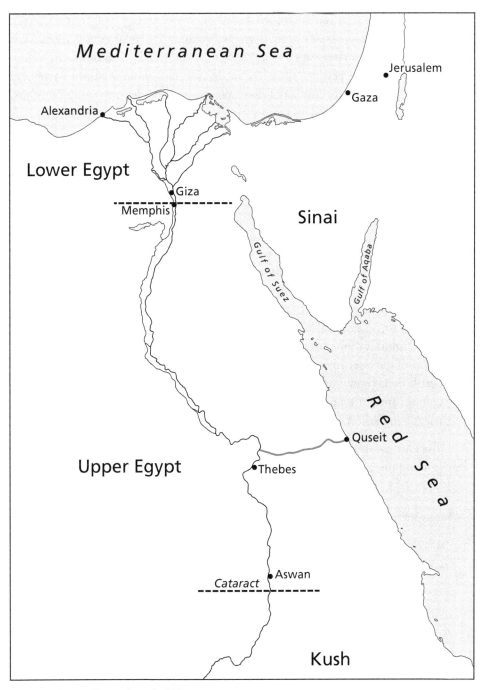

Map 3.2 Ancient Egypt along the Nile.

on the ability of the Egyptians to make use of the natural resources of the Nile valley and in particular the fertile land of the delta. The traditional story of the creation of the empire was the unification of Upper and Lower Egypt[3] by King Mene some time before 3150 BCE. Little archeological record of Mene has been discovered, and it is more likely that the unification was started by the pharaoh Narmer, but by 3150 BCE the first leaders of a unified Egypt established a capital at Ineb Hedj, better known by its Greek name Memphis. This was about 20 km (13 miles) south of modern Cairo on the west bank of the Nile, at the strategically important point where the river begins to spread out into delta. It became a major center for food distribution and industry (Map 3.2).

The Nile flooded annually and this contributed to the power of the empire in a number of ways. The most obvious was agricultural productivity, as the river brought organic material and volcanic silt that renewed the soil every year. The indirect contribution was the promotion of central authority, as dealing with the flooding and then the agricultural production of the growing season required a great deal of coordinated effort. This created a kind of self-boosting system, as increased agricultural production meant a larger population could be supported, which in turn meant greater specialization of skilled work was possible. Those specialized skills, particularly among the rulers and the clerical class, allowed the rulers to organize the workforce to undertake larger projects such as mass irrigation. This increased agricultural production and so on.

Tools such as plows and the use of animal labor increased, and under the central administration, taxes (mostly in the form of food) were collected. With these surpluses, the rulers could undertake major civic projects, especially building temples, palaces, monuments and tombs. In the early period, the biggest projects were the Step Pyramid built for the pharaoh Djoser (c. 2635–2610 BCE), and the Great Pyramid at Giza, built for the pharaoh Khufu (c. 2589–2566 BCE, also known as Cheops), although there is still debate about the precise date of construction and therefore for whom the pyramid was built. The original structure was about 146 m (480 ft) tall, and each side was 230 m (756 ft) long at the base. Its total mass is close to 6 million tonnes. In terms of the amount of work that was required to build the huge structure, historians and engineers have estimated that between 250 and 800 tonnes of stone had to be moved per day to build the pyramid during the reign of Khufu (Figures 3.2 and 3.3).

There have been a great many ideas about how the Egyptians built the Great Pyramid and no clear answers to some basic questions since no images of the construction have ever been found.[4] What we do know is that the Egyptians built the pyramids primarily with paid labor. We may not have construction details, but accounting records have

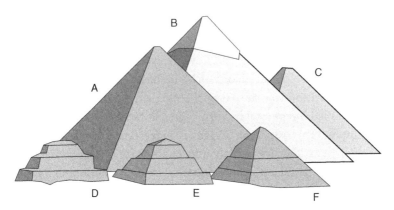

Figure 3.2 Pyramids at Giza. A. Khufu, B. Khafre, C. Menkaure, D. Meresankh III, E. Meritets I, F. Hetepheres I.

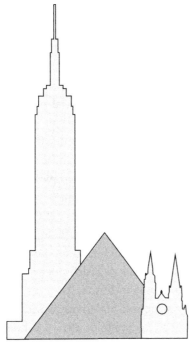

Figure 3.3 Comparing the Great Pyramid (146.7 m; 481 ft) to the Empire State Building (443.2 m; 1,454 ft) and Chartres Cathedral (113 m; 371 ft).

survived telling us that farmers in the off season worked on the site and were paid in grain and onions. In 1990, workers' graves were discovered near the site and the archeologists Zahi Hawass and Mark Lehner suggested that as many as 200,000 skilled workers, divided into two *gangs*, and subdivided into *zaa* or *phyle* of 20,000 each, worked on the construction. It is likely that the number of workers varied greatly at different stages of construction.

The stone for the pyramid was limestone and came from quarries up and down the Nile from Giza and floated to the site on barges. The stone was probably cut using stone mallets, copper chisels and wooden wedges. Holes were made in the limestone and dry wooden wedges driven in. The wedges were then soaked with water and the expanding wood cracked the stone. The largest stones, weighing 60 tonnes, probably came from a quarry near Aswan, some 800 km (500 miles) away. Modern tests have demonstrated that large stones could be moved by relatively small teams of workers of about 6–10 workers per tonne, either by pivoting the stones on balance points or by sliding the stones on greased wooden or stone tracks. The idea that wooden rollers were used might work for small blocks, but such rollers would either be crushed or driven into the ground by larger stones.

The surveying for the construction of the Great Pyramid employed the tools of the age that included water levels, measuring devices, and the knowledge of what we call the Pythagorean relation but is actually an Egyptian mathematical discovery. Simply by putting knots in a loop of cord representing the equal spacing 3, 4 and 5, a perfect right angle could be formed. Relying on a strong knowledge of astronomy, the Great Pyramid was laid out with its sides aligned north–south and east–west with great precision. There is some evidence to suggest that the structure and orientation of the various pyramids in Egypt had astronomical or astrological significance, but this is not uniform for all the pyramids.

The biggest construction mystery was how the workers got the blocks into place as the structure went up. Most archeologists believe that some form of levers and wedges was used to lift the blocks into place, and modern experiments have shown that this could be done. At the lower levels, earth ramps would have been possible, but as the structure got higher, the vast volume of material to make a ramp would have been so huge that creating the ramp would have taken almost as much effort as building the pyramid itself. Further, while some evidence of ramp material exists, it is not in the volume necessary to reach the top of the pyramid. In 1999, architect Jean-Pierre Houdin developed the idea that internal ramps were used to build the upper sections of the Great Pyramid. His theory solves a number of problems and there is evidence to support his claim, including gravimetric scans of the pyramid, but it remains technically possible but not proven.

It is quite possible that future research will shed light on the actual construction methods used to create the Great Pyramid, but what is often overlooked by people hoping to sensationalize its building (even going so far as to suggest that the ancient Egyptians had help from aliens, or that some earlier but wildly advanced civilization built them over a much longer period) is that pyramids were built all over the region, ranging from a few meters in height to hundreds of meters. In particular, the so-called Bent Pyramid, built for the pharaoh Sneferu and located at the royal necropolis of

Dahshur, offers an insight into building of such large-scale projects. Completed about 2596 BCE, the bottom half of the pyramid is set at 52°, but the top slopes at just 43°. This change was likely due to fears that the weight of material was causing the interior to collapse, although it might also have been made to reduce the amount of material needed to finish the project. Either way, the Bent Pyramid suggests that construction problems affected the project and later builders learned from the experience.

Pyramid building was not confined to Egypt, but was a common design found around the world, with pyramids found in South America, across the Fertile Crescent and in Asia. Impressive as many of the giant pyramids were architecturally and in terms of construction, pyramids tell us something more than that the ancients were good engineers. They tell us about societies that could take on such monumental projects. There was a large enough surplus of food and resources to support the armies of workers. There were managers who could oversee the construction, and planners who could design the project. There were skilled workers who could be given instructions and produce the materials needed to order, and further train new workers to undertake the skilled work. At the community level, taking on these big projects, while perhaps not universally accepted as a good thing, was conceivable and thus represented the strength and abilities of the society in more general terms. In other words, impressive as the Great Pyramid is even today, it was the infrastructure of the Egyptian society that made it possible. The pyramid was the end product of a technological system that brought together everyone from the farmers to the pharaohs. That was the great achievement of the ancient empires, not the size of their monuments.

And yet, the Great Pyramid of Giza can also be seen as a symbol of the problem of a technological society. The pyramids and other building projects offered a unifying focus for the empires that built them, providing a symbol of social values and identity. But the pyramids were mausoleums, gigantic tombs with little practical utility. They absorbed resources without the prospect of a functional return, although they served a symbolic role. The Egyptians had the technical skills to make the pyramids, but they were a dead end (literally and figuratively) in terms of technological change. It has even been argued that the pyramids and the Egyptian concentration on death and the afterlife curtailed innovation, since there was little point in improving the present world when the afterlife was so much more important. The drive to create monuments as displays of power seems intimately linked to civilization. When a society becomes powerful enough to undertake such projects they almost always seem to do so.

1 What conditions seemed to favor the establishment of permanent settlements?
2 What lessons does Çatalhöyük teach use about the benefits and problems of permanent settlements?
3 How did the annual flooding of the Nile contribute to the ability of the Egyptians to build the pyramids?

FURTHER READING 61

NOTES

1. There are many variations of this name, including Catal Hayuk and Catal Huyuk.
2. A further Pre-Pottery Neolithic C, 6200–5900 BCE, has been added.
3. The Upper Nile extends from the Cataracts in the south at modern-day Aswan to Zawyet Dahshur (near Cairo). The Lower Nile runs north from Zawyet Dahshur to the delta at the Mediterranean Sea.
4. There are many theories regarding the construction, ranging from the bizarre (aliens) to the technically possible but highly unlikely, such as cast-in-place limestone concrete or the use of massive water lifts.

FURTHER READING

Why and how some humans shifted from hunter-gatherers to farmers is one of the most complex parts of human history. The fascination with this period comes in part from the fact that our ancestors began to undertake monumental building projects. At the heart of this is agriculture, and Serge Svizzero and Clement Tisdell provide an overview of current ideas about the beginning of farming in their article "Theories about the Commencement of Agriculture in Prehistoric Societies: A Critical Evaluation" (2014). Harold Innis offers an important theoretical insight into the reasons that the earliest empires appeared where they did and how they operated, in *Empire and Communication* (1986). The history of early settlements is hotly debated, but an accessible article about Çatalhöyük is Ian Hodder, "Women and Men at Çatalhöyük" (2004). The most popular and the most misrepresented monuments of the ancient world are the pyramids at Giza, but a good starting point is John Romer's *The Great Pyramid: Ancient Egypt Revisited* (2007).

4 The Eastern Age

7000 BCE	Fermented drinks
	Rice cultivation
3600	Silk fabric
3200	Earliest iron objects
3000	Horse harnesses
2200	Iron working
2000	Tea as a medicine
1600	Bronze plow
1100	Civil service examinations
775	Solar eclipse recorded
500	Crop rotation
300	Cast iron
300	Earliest waterwheel
481–221	Horse collar

400	Steel making
200	Gears used with waterwheels
600 CE	Block printing
900	Paper money Gunpowder used for pyrotechnics
1274–95	Voyage of Marco Polo
1288	Oldest existing cannon
1405–33	Seven voyages of Zheng He

🌼 Introduction

In Asia the combination of good agricultural land, growing populations and rich empires led to a period of great creativity. Many of the most historically important inventions, including silk, paper, gunpowder and the horse collar, made China a technological leader. Another factor in the development and spread of technology was the establishment of a well-educated bureaucracy who could develop, spread and use new forms of technology. These characteristics allowed Chinese rulers to take on large projects such as the Great Wall and the Magic Canal, which involved tens of thousands of workers. China's wealth and technological strength was unrivaled, but that contributed to a period of lower technological development since there was no need to innovate and society became more rigid. This was supported by Taoist and Confucian philosophies that favored social stability and discouraged change. As a result, Chinese technologies tended to get bigger and more refined rather than introducing innovations.

🌼 China and East Asia

On September 6, 775 BCE, a solar eclipse was observed and recorded by priests in China. This became the first precise date of which we can be completely certain in history because the orbit of the stars and planets provide an extremely accurate clock and calendar. Although astronomical events were recorded by many early civilizations around the world, it is from China that the oldest written records survive. In Asia, and in China specifically, a great age of civilization made high culture, the systematic investigation of nature, and advanced technology possible to a degree previously unseen.

This era of invention saw devices and practices developed that set the region far ahead of any other in the world. Why it became such a hotbed of invention, and why it then ceased to lead the world in technological development, are questions that historians have investigated for generations (Map 4.1).

The list of inventions that come from China is long and varied. A small sample is shown in Table 4.1.

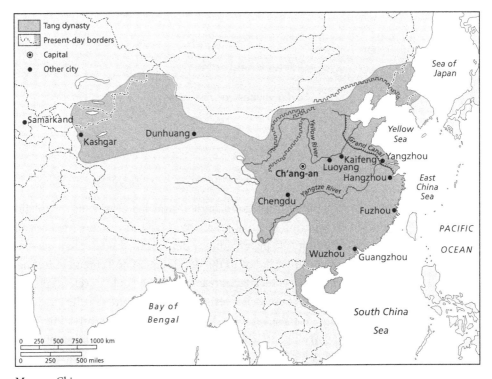

Map 4.1 China.

Table 4.1 Chinese inventions.

Invention	Earliest date known
Fermented drink	7000 BCE
Rice cultivation	7000
Silk	3600
Noodle (millet)	2400
Fork	2400
Bell, pottery	2000
Chopsticks	1200

Table 4.1 (cont.)

Invention	Earliest date known
Kite	500
Wheelbarrow	400
Traction trebuchet	400
Star catalogue	400
Draw loom	400
Natural gas for fuel and light	400
Seed drill, multi-tube	300
Tea, as beverage	200
Seismometer	132 CE
Fishing reel	400
Porcelain	700
Escapement mechanism	700

Important as many of these inventions were individually, they need to be understood in the context of a society that could integrate new tools, systems for work and consumer goods into its existing culture. At the heart of this Asian blossoming was agriculture, on some of the most fertile land in the world. With increased use of iron tools in agriculture, the creation of a surplus and improved methods of storage technology, a great number and variety of people were freed from agricultural labor. While it makes sense that a reliable supply of food would be key to a powerful civilization, one of the main benefits is easily forgotten: simply having more food improved health and provided more energy. More calories made people healthier, stronger and able to take on more work – although as the historian Yuval Noah Hawari notes in his book *Sapiens: A Brief History of Humankind* the archeological record also suggests that modern human physical ailments date from this point. Agricultural productivity was further increased by the introduction around 500 BCE, in both the Indus river region and Mesopotamia, of technology facilitating planting in rows, hoeing weeds and fertilizing crops with manure. In addition, crop rotation began, so that farming did not deplete the soil nutrients. Crop rotation, which would help propel Europe to world power status 2,000 years later, required farmers to think in a new way. Rather than planting all of their land, some was left fallow and then planted with different crops in a seasonal rotation, rather than with the same crop over and over on the same parcel of land. There were a number of benefits to this. Fallow fields and crops such as legumes helped restore vitality to the soil by increasing nitrogen levels and the amount of organic material in the land.

Rotation also helped control crop-specific diseases that were soil-based. By removing an affected crop for one or more seasons, the pathogen had no host and would fade or disappear from the soil before the next rotation.

Horse Power

Physical labor was eased by the use of water buffalo, oxen and horses to carry materials and pull carts and plows. Although oxen were easier to use as draft animals, horses were stronger and could work longer, so they became the preferred farm animal any-where they could be employed. At first, pulling power was simply a matter of putting a rope around the horse's neck, but this was problematic (Figure 4.1). The rope could choke the horse as well as abrading the skin and severely limited the amount of weight any horse could pull. The archeological record suggests that various forms of "throat-girth" harness were concurrently introduced in China and other parts of Asia, starting as early as 3000 BCE. The throat-girth harness placed flat straps across the shoulders and chest of the horse and was connected to the load above the neck. While this spread the load and reduced chaffing, it still pressed against the trachea and constricted the breathing muscles. The harder the horse pulled, the greater the problem.

A variety of archeological evidence suggests that to shift the load away from the neck, the "breast collar" harness was developed between c. 481 BCE and 221 BCE, an era known as the Warring States Period. This shifted the load to the sternum, rather than the neck. It allowed much greater pulling power, although attaching chariots or carts required the straps or poles to go around the sides of the horse. The breast collar harness did, however, offer a more stable harness for mounted riders, and forms of it are still used today. The Warring States era was a time of widespread turmoil, but also the period associated with the shift from bronze to iron in China, and a general increase in technological innovation. This was the period when Sun

Figure 4.1 Stages of horse harnesses. 1. Throat girth consisting of a rope or strap around the horse's throat and chest. 2. Breast strap collar using several straps around the neck and chest. 3. Horse collar using a padded but rigid frame resting on the shoulders.

Zi (Sun Tzu, c. 400–320 BCE), master strategist, is reputed to have lived and written *The Art of War*.

The final stage of development of pulling harness was the introduction of the horse collar some time before 480 CE. While versions of the horse collar may have been in use as early as the third century, no direct evidence exists until the later date. The collar was rigid, and usually of heavy leather or leather over a wooden frame rather than the straps and ropes of the earlier harnesses. It shifted the weight to the shoulders rather than the neck and chest, and because of its rigid construction it transformed the way power was transferred. Instead of pulling on a harness, a horse in a horse collar pushes against the load, allowing the maximum amount of power to be used, including the ability to use gravity, as the horse could lean into the harness. The horse collar also allowed the load to be attached to the horse in different ways such as over the shoulders as well as around the sides or barrel of the horse. Although all forms of harnesses remain in use, the horse collar increases pulling power by 50 percent – a significant gain for the agricultural utility of the plow horse.

Animal power was harnessed to pull loads such as carts and sledges, but was also adapted to draw plows. The concept of the plow was most likely developed in the Fertile Crescent, and there is evidence for plowed fields near Prague, Czech Republic, around 3500 BCE.[1] Plows appear in ancient Egyptian pictures, such as the burial art of Sennedjem, as early as 1200 BCE. By 1000 BCE bronze plows were in use across Asia and Europe, contributing to a rise in agricultural productivity that in turn supported larger populations and greater development of civic society. The combination of the plow and the horse collar helped to transform agriculture. Joseph Needham calculated that a horse could pull about 400 kg (1,100 lb) using a throat girth, but 1,360 kg (3,000 lb) with a horse collar, leading to a significant increase in the amount of work the horse could do (see Needham 1986: vol. 2). Farm production increased, more land could be farmed by the same number of farmers, and the utility of horses made them even more valuable. Horses soon replaced oxen in most parts of the world since they could do more and work longer. The major exception to the use of the horse in agriculture was in rice paddies or wetland farming. In these conditions water buffalo or human power was used.

Moving materials, especially heavy materials such a stone, led to the invention first of sledges (essentially logs or boards dragged along the ground) and later of wheeled carts. It is not clear where the wheel first appeared, with evidence discovered in Mesopotamia, the Caucasus region between the Black Sea and the Caspian Sea, and the Ukraine for use of the wheel around 4000 BCE. The earliest dated artifact is the Bronocice pot found in Poland that appears to show a four-wheeled cart (Anthony 2007). It dates to about 3500 BCE. The wheel spread almost everywhere there was agriculture, with the notable exception of North and South America, where it appears to have been independently invented by never used on a large scale.

❧ Metallurgy and New Tools

One of the greatest areas of innovation was in metallurgy. By 1600 BCE, bronze work was becoming very sophisticated. This was due, in large part, to the application of bellows to heat the molten metal to high temperatures. In turn, this meant that the smiths could cast more objects more quickly and with finer detail. These items, especially bronze weapons, were highly desirable trade items, and bronze weapons from China had made their way to the Middle East by the fourteenth century BCE. Casting larger and larger objects such as temple bells and vessels became widespread from around 1200 BCE.

By 310 BCE, blacksmiths had a double-acting bellows to produce a continuous stream of air that heated their furnaces and smelters. They had also begun to use coal rather than wood or charcoal for their furnaces, leading to higher and more consistent temperatures. These two events contributed to the invention of cast iron around 300 BCE. This required a minimum temperature of 1,200°C to melt, significantly higher than bronze at 915°C. Although iron and even steel had been produced in small amounts going back to 1500 BCE, it had all been forge worked, meaning that the ore was only heated to the point that it could be shaped by hammering. With the greater heat produced by coal and the new bellows, iron could be heated to its liquefaction point (at about 1,400°C), making it possible to cast it in molds. Casting was expensive in terms of fuel and equipment, but the superior quality of iron for tools, weapons, nails and spikes made it worth the high cost of production. The early types of cast iron were very brittle due to the high carbon content, but years of experimentation with the heating and cooling techniques, as well as the development of techniques to clean out impurities, led to a malleable form of cast iron that made it even more useful (Table 4.2).

The iron industry grew to be so important in China that in 119 BCE the Han dynasty effectively nationalized production, bringing it under centralized control.

Table 4.2 Metal working.

Annealing	Heating metal to a high temperature. When the metal becomes malleable it is allowed to cool slowly, usually at room temperature. This alters the physical properties of the crystalline structure. It reduces internal stress and makes material more ductile and flexible. It is easier to shape without fracturing, but also less hard.
Hardening	Heating metal to a high temperature. When it becomes malleable it is quickly cooled (often called quenching), often in water or oil. This makes the physical structure of the metal more rigid, increasing it hardness (and ability to take a sharp edge), but decreasing flexibility and increasing the chances of fracturing.
Tempering	After hardening, metal is heated and allowed to cool slowly (annealing). A cycle of heating and fast or slow cooling allows smiths to control the balance of hardness and flexibility.

The utility of cast iron as a construction material was considerable and by 10 CE complete structures could be made from it. Perhaps the greatest demonstration of the industrial capacity of the iron workers of China was the great iron column ordered by the Empress Wu Zetian in 695 CE. Made in several parts, its total weight was about 700 tonnes. The largest single casting was the Iron Lion of Cangzhou (also known as the Great Lion of Tsang-chou and Sea Guard Howler), cast in 953 CE to commemorate the Chinese victory over the Tartars. The statue was cast using piece molding, a technique borrowed from bronze casting. The lion was first modeled in clay, and when the clay dried it was covered in a second layer of clay. This formed the exterior mold with all the details from the original figure. While it was still wet, the outer layer was cut into sections and removed to dry. The artists carefully shaved down the lion model so that when the outer layer was put back around it, there would be a gap or void between the inner and outer sections. The molten iron was poured in stages, filling the gap. It weighed 40 tonnes and it must have taken over a week for the mold to cool enough for the clay shell to be removed.

The peak of Chinese metallurgy in this period was the discovery around 400 CE of a method for making steel, an alloy of iron with carbon, as well as other trace elements such as nickel. The earliest steel was produced by combining cast iron with wrought iron. Steel is harder than iron and can hold a much sharper edge. Steel objects made using naturally occurring alloys had been made as early as 500 BCE in India, so the qualities of steel were known in Asia and the Middle East before blacksmiths worked out a method of production. The manufacturing of this natural form of steel, known as Damascus steel in Europe, diminished over time when the supply of ore ran out. Running out of natural resources has been a theme throughout human history, but it has also been a prompt to invention. Efforts to duplicate the qualities of this material led to innovations in metallurgy, particularly by the steel producers of the Industrial Revolution.

Although Chinese steel was very good, the ultimate edged weapons were made in Japan by the smiths who made samurai swords. Ancient swords have been found in Japan from before 900 CE, but the iconic curved "katana" began to appear in the Kamakura era (1185–1333). The new type of sword was longer and slightly curved to facilitate fighting on horseback and was a response to the Mongol invasions of 1274 and 1281. The older swords, although well made, were too light to cut the leather armor used by the invaders. This process demonstrated a brilliant understanding of the properties of metals and metallurgy, since the swords were a combination of iron and steel. The steel allowed a sword to be rigid and hold an edge, while the more malleable iron gave it flexibility. By a process of repeated folding, the metals were fused together in thousands of layers. The repeated beating out drove out impurities that would weaken the metal, and also led to the formation of a very fine crystalline structure within the metal. The smaller the crystals, the less prone to fracturing. Finishing the swords was a careful process of quick and slow cooling that contributed to the strength and resilience of the weapon. The finished product was so sharp that one blow could slice through a man diagonally from the shoulder to the hip.

Waterwheels

Although the use of human and animal power was common in China, waterwheel technology was one of the key developments that made the mass production of cast iron and other products possible. The earliest waterwheels were not designed to deliver power, but to raise water for irrigation. By the fourth century BCE, the ancient Egyptians had invented a wheel with pots or buckets on it that allowed people to lift water from a stream or river to irrigation ditches using the power of the flowing water.

It is not clear when the idea of using the motion to do other kinds of work was discovered, but the first water mills were direct-drive devices used to grind grain. These appear to have been independently invented in Greece around the beginning of the third century BCE and in China mills were in operation before the first century CE. The early grain mills used a horizontal wheel that drove a shaft and turned a millstone (Figure 4.2).

Sometime around 200 BCE, gears were added. Gears allowed motion to change direction, so that a vertical motion could be turned into a horizontal motion, and as the principles of gears became more complex, the speed of rotation could be speeded up

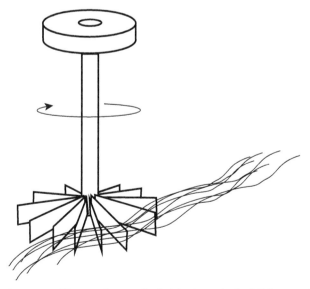

Figure 4.2 Horizontal waterwheel. A horizontal wheel did not need gears, but its speed was limited to the speed of the water.

or slowed down using different sizes of gears. A circular motion could be turned into back and forth motion using an offset crank. This configuration could be used to run saws or bellows. The use of water power, as well as wind power where it was suitable, had become common in China and across Asia by 200 BCE.

In part because of the economic wealth generated in this period, Chinese society developed strata of government officials, courtiers and scholars. Thus the innovations in manufacturing were accompanied by a blossoming of culture and scholarship. Chinese scholars developed interests in the collection and classification of the things they saw in the world around them, along with a great interest in the strange, novel and magical. Alchemy became popular and it helped to foster an interest in matter theory or ideas about the types and behavior of matter in physical objects. This, in turn, contributed to the development of chemical industries such as the production of acids, gunpowder and dyes.

Another object with "magical" properties was the lodestone, a naturally occurring magnet (often a fragment of a meteorite). Lodestones had intrigued people for generations. The oldest reference is from Thales of Miletus in the sixth century BCE. Lodestones seem to magically attract iron objects or move without being touched and were likely used for divination, but the practical application of a lodestone to determine direction is not completely clear. A lodestone used to determine direction seems to be noted in the *Book of the Devil Valley Master* (c. 300 BCE). It was called a "south-pointer" because that was the direction considered most important to know. The use of the lodestone as a compass by Chinese mariners was mentioned in Zhu Yu's *Pingchow Table Talks* (c. 1111 CE) but it may have been in use as early as 900 CE.

Time keeping and record keeping were important to Asian civilizations. People from India to the Pacific islands developed impressive astronomical observations, learning to track time and navigate by the stars, the sun, the moon and planets. A strong belief in astrology also helped motivate this interest in the heavens. Shadow clocks had been known in Egypt as early as 1500 BCE, and the sundial was extensively used in China by 600 BCE. This was significant both for its utility (at least when it was sunny) and for the concept of time that the device embodied.

In addition to astronomical methods to keep track of time, the Chinese used water clocks (also known as "clepsydra"). The earliest water clocks were simple devices consisting of two vessels, usually ceramic bowls or pots, placed one above the other. Water flowed out of a small hole from the upper pot into the lower pot. Examples of this type of water clock come to us from Egypt and Babylon around the sixteenth century BCE. In China, Joseph Needham found evidence of water clocks as early as the sixth century BCE. By the eleventh century CE, Chinese water clocks had become elaborate mechanical devices using waterwheels, gears, and escapements (which controlled the speed a gear turned) to operate dials, turn armillary spheres and ring chimes. Time was translated from the great sweeps of the heavenly bodies to the passing of the hours as marked out on the sundial and the water clock. The measurement of time was closely linked to the development of court and government life, as activities were segmented and controlled by time.

❧ Chinese Bureaucracy and Paper

The development of the Chinese imperial bureaucracy was one of the key ingredients in the rising sophistication and use of technology in China, allowing the Chinese to conceive and undertake complex projects. Chinese society was extremely hierarchical; with power descending from the emperor to the nobles and down to the farmers, there was a constant struggle for the degree and autonomy of power of each group. This struggle sometimes resulted in military conflict, but often it was a more subtle struggle with court intrigue, shifting political support and the pursuit of status driving social activity.

Over time, particularly in the Tang (618–907) and Song dynasties (960–1279), there grew up at the imperial court an increasingly powerful group of professional advisors and ministers, and working for them were legions of lesser bureaucrats. If the government hoped to control the vast population and utilize the power of the people and the resources of the country, it had to develop systems to record vast amounts of information, from ministerial reports to tax rolls. In the early days of the bureaucracy prior to the Warring States period (475–403 BCE), records were kept on a variety of materials such as silk, wood, hide and bamboo strips stitched together to form scrolls. These materials were expensive to produce, so as the demands for record keeping grew, so did the cost of keeping those records. Around 100 CE, paper was introduced for writing. It was made from a variety of materials, with mulberry bark a favorite choice as it made a very pliable material without a great deal of processing. Paper was not unique to China. In Fiji and Samoa, tapa cloth, made by beating mulberry bark, was used in prehistoric times and continues to be made today. It was in China, however, that the widespread production of paper as a writing material was pioneered.

The production of written records and communications in the Chinese bureaucracy took on almost industrial dimensions. Imperial announcements, for example, had to be sent to hundreds and sometimes thousands of locations. During the Tang dynasty (618–907 CE) many, many copies of the *Kaiyuan Za Bao* ("Bulletin of the Court") were handwritten on silk and circulated to government officials. This meant an army of scribes had to reproduce documents. Scribes were also responsible for copying any other texts, from poetry to tax receipts. Although it might seem that printing should have been an obvious solution to the problem of producing government texts, it was a technology that was not adopted for this purpose until long after it was technically possible to use printing. The number of people who were literate was limited, so the skill associated with writing was highly specialized and highly prized. Even the most junior scribe was undertaking a skilled job, while at the highest intellectual and social level writing was an art form, not just a utilitarian activity. A great deal of attention was paid to the formulation of ink (hide glue, carbon black and bone char) and the manufacturing of brushes. The ink, made liquid by the addition of water, was opaque, stable and long lasting. The earliest archeological discovery of a Chinese calligraphy brush is from c. 300 BCE, found in a tomb near Changsha. Even today, people in China can be seen practicing their calligraphy in public parks, using water on the sidewalks. As the needs of the government grew, the number of scribes grew accordingly, so the needs of the government were met and there was little impetus for change.

In a sense, printing was an extension of ancient art and craft traditions. In textile art, objects such as hands, leaves or other common objects, were colored and then pressed against cloth. Later, people began to carve patterns in wood, clay, leather or stone and use those to print fabric. Another form of printing used cylinder seals to transfer an imprint to soft clay. These were created in Mesopotamia as early as 3000 BCE. Various kinds of maker marks (a symbol printed or stamped onto an object by the maker) were also known in antiquity, including the Chinese chop or signature stamp. It was not a big step from a stamp for a maker mark to using stamps for printing. Artists and calligraphers developed wood block printing of images, and this became a form of reproduction around 200 CE.

The earliest use of woodblock printing of text dates to around 600 CE, when whole pages were carved and printed from a single piece of wood. The oldest surviving example of such printing is a Buddhist scroll from around 704 CE. The oldest dated printed text is the *Diamond Sutra*, produced in 868 CE. The 5 m (16 ft) scroll is sometimes referred to as the first printed book because it was the first extended piece of printed literature. The combination of cheap paper and the skills of woodcarvers came together to create a print industry in China that led to the first example of mass communication.

Block printing was (and continues to be) used for textiles and artistic work, but it was awkward and time consuming to use for text printing. A solution to the problem of text printing was the introduction of movable type printing sometime around 1045 CE. The creation of movable type printing is credited to Bi Sheng (990–1051 BCE). The story of Bi Sheng comes from a single source, *Writings Beside the Meng Creek* (or its more poetic title, *Dream Pool Essays*) c. 1088 BCE. Written by Shen Kuo (1031–95), an important scholar and statesman, it tells of Bi Sheng, a commoner, who made characters in soft clay, then fired them and assembled the clay blocks on an iron frame, gluing them into place with resin and wax. The technical detail suggests that Shen saw such printing in operation.

Although clay was an easy medium for the creation of characters, it was unsuitable for mass printing because it could be irregular and disintegrated. Around 1300 CE, carved wood type was introduced in China, while the earliest use of metal type that exists today dates from Korea from 1377 CE, and bronze type characters were in use in China by 1490 CE. The characters were hand-carved.

The use of movable type did not have as big an impact on Chinese society as it did in Europe. There were some limitations that were never really overcome in the Chinese version. The number of characters in the written Chinese language, even in simplified vocabulary, was large, running to between 2,000 and 3,000 character blocks. In the early period, the use of clay or carved wood limited the speed at which blocks could be produced since they had to be hand-carved, and it also meant that they had a shorter life, wearing out quickly. And perhaps as significant as the technical issues, there was a much stronger scribal tradition to overcome than in the European version of the printing story in Chapter 6. In China, scribes were part of the imperial bureaucracy and had responsibilities beyond just transcribing and acting as secretaries. Although printing would become an important art form and industry throughout Asia, it did not contribute to the transformation of Chinese society in the way that it is usually argued to have done in Europe.

At the intersection of the technologies of printing and paper was the development of paper money in China around 800 CE. Starting as bank drafts used to conduct long-distance exchanges without having to transport large quantities of bullion, after the government took over the issuing of drafts in 812 CE it developed into a medium of exchange around 900 CE. In 1107 CE, multi-color printing was introduced to make paper money harder to counterfeit, setting off a kind of printing arms race between counterfeiters and governments that continues around the world to this day as the people issuing paper money introduce new features (such as microprint and watermarks) to make counterfeiting more difficult.

The government of China required local administrators, regional administrators and national ministers. The bureaucracy of the centralized government was very large since almost all important decisions and policies were decided at the court of the emperor. The court was not just about government; it also functioned as the spiritual and cultural hub of China. As such, judges, spies, engineers, generals, astrologers, teachers, priests, artists, doctors, scribes and the occasional alchemist were part of the court. It was an extraordinary concentration of talents, and such a concentration of power tends to look for ways to demonstrate the power of the bureaucracy to the wider society and, by doing so, demonstrate their utility to the court and the emperor.

By 300 BCE the development of agriculture and industry in China was so significant that the governments could undertake massive building projects. During the Warring States period (c. 475–221 BCE), many of the different states had constructed massive earthen works and walls to protect themselves. When Qin Shi Huang conquered all the states and unified China in 221 BCE, he declared himself emperor. To deter revolt, he ordered the removal of the internal walls and barricades that had been built by the various warring states. Quin Shi Huang was unable to subdue the Xiongnu people in the north and the threat of raids and invasion led him to order the building of a wall and fortifications along the empire's northern frontier. The Qin wall was a precursor to the Great Wall of the later Ming dynasty (1368–1644 CE). Although it is not clear today what the extent of the early wall was, since much of it has been worn away, taken down for its materials or overbuilt by later wall works, it was a huge project running hundreds of kilometers and requiring the efforts of thousands of workers.

In 214 BCE, Qin Shi Huang, ordered the construction of the Lingqu or "Magic Canal." Qin used the canal to send troops and supplies south to conquer the region. Lingqu was a 32 km (20 mile) link between the northern Xiang river, a tributary of the Yangtze, and the Li river that flowed south to Guangzhou. To overcome the differences in altitude, the canal used flash locks which were channels that could be opened or closed beside rapids. Boats going down rode the current, while boats going upstream had to be pulled by ropes against the current. The canal was extended to 145 km (90 miles) connecting the Ch'ang-an river to the Yellow river in 215 BCE. These massive civil engineering projects served practical purposes such as moving troops and food, but they also provided irrigation and a communications network, speeding up the flow of information and allowing greater centralized control of the empire, in much the same way as the roads of imperial Rome aided Roman emperors.

Gunpowder

One of the most powerful inventions, literally and figuratively, was gunpowder. The potentially dangerous results of mixing sulfur, charcoal and saltpeter were discovered by Chinese iatrochemists, or what we might call alchemists, in the course of their efforts to produce elixirs of immortality, sometime before 900 CE. By 900, gunpowder was being used for pyrotechnics and it was probably around this time that its explosive power was discovered. In 1040, the Chinese scholar Zeng Gongliang published recipes for three forms of gunpowder. By 1150, it was being used for rockets. The first record of gunpowder bombs may be from the Battle of Caishi in 1161 CE, and bombs were certainly in use by the time of the Mongol invasion of Japan in 1274 CE.

At about this time, knowledge of gunpowder traveled west to the Islamic world, and in 1249 CE gunpowder was used by Islamic forces against European Crusaders in Palestine. Thirteenth-century China witnessed the era of Mongol conquest, and under the Mongols there was a swift development of gunpowder weapons. One of the earliest uses of gunpowder as a projectile launcher was a "fire-lance" that consisted of a tube attached to a lance or long arrow. The tube was packed with gunpowder, lead shot and other projectiles. When it was set off by lighting the powder, it spouted flame and shot out the projectiles as it burned. This was more like fireworks than a gun, however. The first true gun was developed around 1280, and the oldest existing cannon comes from

1288. It consists of a bronze barrel with a touch-hole toward the back of the barrel. Found in Manchuria, it was probably used to propel a large bolt or arrow rather than a cannonball.

Invisible Technology in China

The ability of the Chinese to invent new devices and methods of production was impressive, but more impressive was the ability to integrate and spread such new technology, and to utilize the resulting benefits to undertake very large projects such as the Grand Canal and the Great Wall. The organization of mass agriculture and the building of large cities, however, had the greater impact on society. At the center of this social system was the imperial bureaucracy. Yet although there was often a close link between the bureaucrats and the aristocracy, the strong centralized power of the emperor meant that the government was made up primarily of what we might now think of as civil servants owing their allegiance to the emperor, rather than members of the aristocracy filling the high posts in government. This civil service was responsible for overseeing every aspect of Chinese life from agriculture to law to rituals and celebrations. Technological innovation for much of China's history was thus brought into use by members of the bureaucracy (who often had the time and education to work on such engineering tasks), or had to be approved by bureaucrats.

A particularly successful form of invisible technology was the introduction of the civil service examinations to promote competence in the growing bureaucracy of the imperial court. Starting in 124 BCE, Emperor Wu began to examine candidates for office. Prior to the examinations government posts were usually awarded to people recommended by high officials, important families and members of the imperial court, but this did not always guarantee that those recommended were able to do the job. By testing candidates, the most able could be chosen and the emperor could discourage the dynastic ambitions of the nobles. In 605 CE, Emperor Taizong (T'ang T'ai Tsung) initiated a formal examination system that would last, in various forms, until 1905. Successfully passing the examination meant social respect and the chance of a good posting within the civil service.

The foundation of the civil service examination was a knowledge of the Classics, a long list of literary works including poetry, essays, histories, military strategy, Confucian philosophy, law, court life and rites. The candidates were required to have a detailed knowledge of the texts, and be able to write erudite commentaries on the texts in skillful calligraphy. Prior to the institutional examination system most formal teaching was done by tutors in the homes of wealthy families. As the examinations became a way to gain status and employment, teachers became more valuable and schools offering to train boys for the examinations sprang up across China.

In theory, any male in China could sit the examinations, with a few exceptions such as executioners and actors, who were excluded by their lowly social status. In some periods, the merchant class was excluded, but that was not always the case, nor was it easy to enforce. In practice, the cost of study was generally so high that most of the candidates came from the landed gentry who could afford to hire tutors, send children to schools, purchase the texts and devote the time to study. Yet despite the financial constraints, some successful candidates came from the poorest groups and the examinations thereby offered the hope for quick advancement if a boy was smart and dedicated enough.

The form of the examinations changed over time, but in general there were local examinations held in prefecture cities at about eighteen-month intervals. The tests were grueling, lasting several days. Those who passed were not guaranteed posts in the government, but could proceed to the next level of testing held every three years at provincial capitals. To give a sense of the scale of these examinations, at their height, in the sixteenth century, the examination grounds at just one provincial capital at Guangzhou covered 6.5 hectares (16 acres) and consisted of 8,653 individual cells or test rooms.

The successful candidates were sent to the capital for the highest-level examination. Pass rates were on average around 5 percent, while pass rates for the highest examination, the jinshi degree, could be as low as 1 or 2 percent. Those who passed could expect to be invited to fill important government posts as they became vacant. If the civil service did not come calling, there were many other opportunities for the celebrated degree winners such as teaching, tutoring or working for noble families.

The examinations identified boys and men with a particular kind of intellectual capacity and the drive to master complex material. It placed a great value on scholarship, and raised the status of the scholar to a very high level. Family celebrations of a candidate passing even the lowest examination level often rivaled weddings. The examinations allowed some degree of social mobility and brought people from around the empire together. They also meant that the members of the bureaucracy shared a unified intellectual heritage and had passed a difficult trial. To avoid local influences, successful candidates were rarely posted to their home town or region. This helped to ensure loyalty to the central government since bureaucrats had more in common with other government officials than they did with the local population. The examinations did not completely end favoritism or corruption in government, and exam cheating was an ongoing problem, but it was a far more equitable system than had existed before.

The power of the bureaucracy both was encouraged and limited the spread of technology. Many of the greatest inventions of the empire came from the scholarly class. The quick spreading of ideas that solved problems that the civil servants needed to solve was based in part on the ability to communicate those solutions and the acceptance of the information by administrators who trusted the source. But the intellectual power of the civil service was also a problem since it placed such a high value on replicating orthodox ideas. Scholars were expected to have a broad range of knowledge running from poetry to astronomy, but there was little interest in the creation of new knowledge. This often meant that civil servants were open to a broad range of ideas from many sources, but it could also mean that ideas that did not fit easily into the strongly Confucian philosophy at the heart of Chinese scholarship were ignored or rejected. The test system reinforced the idea that stability and continuity were valued more highly than innovation. As a result, some of the most impressive accomplishments in Chinese history were based on a massive application of existing technology rather than the introduction of new solutions. Impressive as the canals and the Great Wall are, they were built using massive numbers of workers using tools that were unchanged in centuries. It was, in political and technological terms, easier to maintain the existing system for undertaking large projects than to develop new tools and systems. When the emperor and the court were interested in new ideas, exploration and innovation, and not distracted by wars, the bureaucracy facilitated invention, but when the leadership was conservative, innovation was limited. From the point of view of the emperor and the court, technology was at a pinnacle, since anything they ordered to be built, manufactured or supplied was produced as ordered.

The power of a strong, unified and well-educated bureaucracy was not lost on other nations. The Koreans borrowed the concept of the civil service examinations in 958 CE, and centuries later the Prussians introduced a version of it in the 1800s. In England, the East India Company established a college in 1806 explicitly to train administrators for the British Empire based on the Chinese system. In 1854, the Northcote-Trevelyan report to parliament began the process that established the civil service examination system for the British government, and many of the Commonwealth countries followed the British practice. Indeed, the Indian government continued this tradition to staff the Indian civil service after independence in 1947. In various forms, most governments in the industrial world continue to have civil service examinations, although as higher education has become more common, the necessity of such examinations for admission to the ranks of the bureaucracy has declined.

Trade and Exploration

Long-distance trade between China, India, the Middle East and Europe existed as early as the time of the Roman Empire, and probably earlier. The desire of people to gain access to the goods, such as silk, manufactured in China and other parts of Asia was one of the great motivators of exploration. In large part because of later European domination of trade, there was a sense in the west that the Chinese were not interested in exploration. Europeans learned the story of Marco Polo (1254–c. 1324) and his voyage to China that was made around 1274 and lasted until his return to Venice in 1295. Marco Polo's story *Il Milione* or *The Travels of Marco Polo* was wildly popular, circulating in various forms and in different languages in manuscript long before the era of the printing press in Europe. Even if scholars have raised questions about whether he actually made the voyage he claimed, his stories shaped the image of far away and exotic Cathay for generations of Europeans.

Yet in fact there were important Chinese voyages of discovery. In particular, the seven voyages of Zheng He (Cheng Ho, c. 1371–1435) demonstrated the power of Chinese technology (Table 4.3). Zheng He came from a family of travelers who were of the predominantly Muslim Hui people and his father was reported to have made the pilgrimage to Mecca. The Ming emperors Zhu Di and his grandson Zhu Zhanji directed Zheng He to take a fleet into the South China Sea and the Indian Ocean. The voyages were multipurpose, including the suppression of pirate activity around Sumatra and spreading Chinese influence in places like Sri Lanka, but the main purpose was exploration and discovery. The voyages took place between 1405 and 1433, and were massive in scale. More than a hundred ships, including some with keels over 100 m (330 ft) long, and as many as 28,000 crew members made the voyages that took the fleet to India, Indonesia, Thailand and Sri Lanka, then across to the Arabian peninsula and down the coast of East Africa, perhaps as far as the Cape of Good Hope. Zheng presented gifts of gold, silver, porcelain, silk and other trade goods and took home exotic animals such as giraffes and ostriches, and supplies of ivory. He also brought envoys from as many as thirty governments to the Chinese court. The voyages promoted trade and Chinese influence, in part by relocating a number of Chinese

Table 4.3 The seven voyages of Zheng He.

1	1405–7	Champa, Java, Palembang, Malacca, Aru, Sumatra, Lambri, Sri Lanka, Kollam, Cochin, Calicut
2	1407–9	Champa, Java, Thailand, Cochin, Sri Lanka
3	1409–11	Champa, Java, Malacca, Sumatra, Ceylon, Quilon, Cochin, Calicut, Siam, Lambri, Kaya, Coimbatore, Puttanpur
4	1413–15	Champa, Java, Palembang, Malacca, Sumatra, Sri Lanka, Cochin, Calicut, Kayal, Pahang, Kelantan, Aru, Lambri, Hormuz, Maldives, Mogadishu, Brawa, Malindi, Aden, Muscat, Dhufar
5	1416–19	Champa, Pahang, Java, Malacca, Sumatra, Lambri, Sri Lanka, Sharwayn, Cochin, Calicut, Hormuz, Maldives, Mogadishu, Brawa, Malindi, Aden
6	1421–2	Hormuz, Arabian peninsula, East Africa, Madagascar
7	1430–3	Champa, Java, Palembang, Malacca, Sumatra, Sri Lanka, Calicut, Hormuz

Muslims to Malacca and a number of other places, where they established trade routes and created centers of Islamic culture in the eastern seas. One significant result of the voyages was the creation of a series of maps, known as the Mao Kun maps, which are some of the earliest navigational charts ever made, and also some of the earliest maps ever printed.

The last of the voyages marked the end of Chinese overseas exploration, and has been seen as a turning point in Chinese history. Concern about foreign influence, the disruptive effect of trade and the influx of people into China was heightened by conflict with the Mongolian people in the north. Resources that had been used for naval work in the south were redirected to military use.

Historians have speculated that the turning away from naval exploration limited China's ability to develop maritime technology and thus made them more vulnerable to the eventual incursions of the Europeans. This in turn has been used to argue that the ultra-conservative Confucian bureaucrats who supported the *hai jin* or "Sea Ban" laws that limited or outlawed maritime shipping brought an end to Chinese innovation starting in the sixteenth century. At a very basic level, China did become less innovative in this period, but the reasons were more complex than a combination of xenophobia and conservative philosophy. War, the diversion of resources and managerial skill to other projects such as building the Great Wall, and overseeing the growing population were far more significant than abandoning overseas exploration. Bringing home giraffes and ostriches provided novelty, but did not impress the scholars or suggest the need for technological change. If anything, the voyages of Zheng He confirmed the superiority of Chinese technology and culture.

When the Portuguese and other European naval forces arrived in Asia in the sixteenth century, they found that their ships and weapons were militarily superior to most of the maritime craft of the Islamic states around Africa, where only the Red Sea was preserved from European encroachment by sheer weight of numbers of opposing forces. The European fleets were stronger than the Indian naval forces, and the Chinese lacked any significant naval presence to challenge the newcomers.

Asian officials were perhaps right to worry about the potential problems created by contact with outsiders. When the Portuguese landed in Japan in 1543, they introduced firearms and contributed to a change in Japanese society, demonstrated in the Shimabara Rebellion of 1637–8, when peasants, many of them Christian converts, rebelled against high taxation and religious persecution. The rebellion was put down, but Japan, like China, imposed restrictions on foreigners and their own people in an attempt to limit the influence of foreigners. Fearing (with good reason in many cases) plots to increase European power, and the disruption of social order that European ideas and technology would bring, Japan entered a 200-year period of isolationism.

The Lessons of Chinese Technology

China's technological history suggests some important points about the place of technology in society:

1 The vast range of tools, devices and technological systems that were developed in China were built on a foundation of agriculture that could support many non-agricultural workers. The life of the average farmer was hard, but offered security and a system to deal with problems.

2 Innovation and a bureaucracy that is reliable and well educated tended to reinforce each other. Innovation could solve problems, and solutions could be transmitted easily and built upon by others who shared a common background and intellectual training. There were large numbers of people with the skills and training to maintain the political system and the technology.

3 Lack of competitive pressure tends to reduce the need to innovate. When the emperors of China were convinced that they lived in the best possible world, there was less interest in innovation. Consequently, the scale of production was more important than the development of new products or systems.

4 Too much competition, particularly in the form of war or civil unrest, tended to suppress innovation as the central authorities directed resources (both material and intellectual) to the defense of the state.

Ancient China was one the great cradles of technological development. China brought us inventions, ranging from the completely practical such as new types of plows to the exotic such as mechanized water clocks. The Chinese also created a powerful invisible technology with its civil service exam and mass bureaucracy. The age of innovation in the East did not end abruptly or completely, but invention became less important in an era when the material needs of the rulers were filled in the often elegant and massive scales made possible by the Chinese bureaucracy and industry. Over time, the civil service examination promoted men with a high level of intellectual ability, but it symbolized the drive for continuity and the rejection of ideas or inventions that might disrupt social order. Security and stability became the objectives of the society. The wonders of the East would, however, become a magnet to other people. They, in turn, would have to solve technological problems to exploit the resources they sought.

1 How did the bureaucracy of imperial China both aid and restrict the development of technology?

2 What do the voyages of Zheng He tell us about the attitude of the imperial court to the outside world?

3 In what ways were the Spice Road and the Silk Road pathways for technology as well as trade?

NOTE

1. Institute of Archaeology of the CAS, Prague, www.arup.cas.cz/?p=12517.

FURTHER READING

Asia, and China in particular, became the great cradle of invention shortly after the development of agriculture and settlements. A very accessible introduction to Chinese history is Patricia Buckley Ebrey's *The Cambridge Illustrated History of China* (1996), but the most detailed investigation of science and technology available in English remains the volumes in Joseph Needham's series *Science and Civilisation in China* (1954–86). India is less well documented for Western scholars, but the collection of articles edited by Manabendu Banerjee and Bijoya Goswami in *Science and Technology in Ancient India* (1994) offers some introductory material. Although it covers a later period, *The New Cambridge History of India* (1987–2005) offers some material on technology. Other areas of Asia in this period are even less accessible and new work in Asia will undoubtedly change our understanding in the future.

5 The Mediterranean World to the Islamic Renaissance

509 BCE	Roman Republic
334–323	Alexander the Great begins his conquests but dies
c. 280	Library of Alexandria founded
c. 287–c. 212	Archimedes of Syracuse
214	Second Punic War; Rome attacks Syracuse
70–80 CE	Roman Coliseum
180	Barbegal mills built
200	Wootz steel (Damascus steel) appears in Middle East
405	Division of Roman Empire
476	End of Roman Empire in western Europe
622	Beginning of Islam
630	Beginning of Hajj (pilgrimage to Mecca)
700–1200	Islamic Renaissance
762	City of Baghdad founded

c. 813	House of Wisdom operating in Baghdad
850	*Kitab al-Hiyal* (*Book of Ingenious Devices* or *Book of Tricks*) by Muhammad Ahmad and Hasan bin Musa ibn Shakir
1095–9	First Crusade captures Jerusalem
1258	Baghdad sacked by Mongols

❧ Introduction

Around the Mediterranean basin a series of empires rose and fell. Part of the strength of the new empires was their ability to absorb technology from other societies and develop new technologies to overcome social and natural problems. The Greeks took over Egypt and expanded on the physical and mechanical understanding of the world, but it was the Romans who created a massive empire based on technological innovations. In particular, roads, a well-trained military and a centralized government allowed Rome to control a vast empire. The wealth of Rome was expressed in civil engineering projects such as aqueducts, temples and large buildings. They introduced the use of the arch and concrete for their building projects. After the fall of the western Roman Empire, Islamic empires grew up and became wealthy by controlling trade between Africa, Europe and Asia. Religion and social mobility spread new technologies east and west, leading to the Islamic Renaissance. In particular, the Hajj or pilgrimage to Mecca, meant that people from China to Spain came together, bringing their knowledge and ideas with them, while the Zakat or requirement of charity contributed to civic projects such as the building of hospitals and schools.

❧ The Mediterranean World to the Islamic Renaissance

The eastern region of the Mediterranean, including the area of the Fertile Crescent, was the birthplace of Western civilization and was a hotbed of cultural development for generations (Map 5.1). A series of empires rose and fell from the dawning of civilization in the region. Ancient cities such as Ur, Eridu and 'Ubaid tell us about the transformation of life from the Neolithic to the early agrarian settlement and then to functioning city-states and empires. This cycle was interrupted by the rise of a new locus of power on the northern side of the Mediterranean basin, as first city-states in a large geographical area we now know as Greece and then Rome rose to imperial power

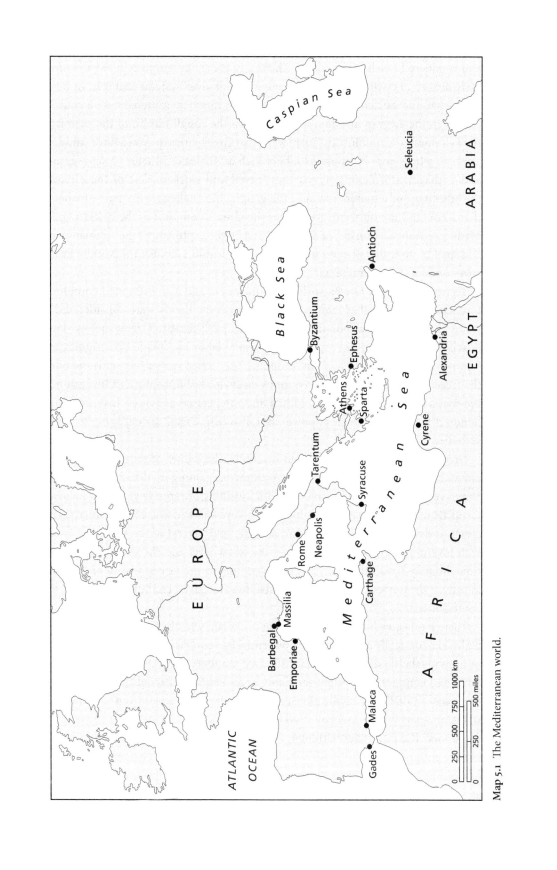

Map 5.1 The Mediterranean world.

and conquered much of the region. The rise of these new empires broke away from the early history of civilization being based on major river systems and was, in hindsight, somewhat unexpected. The Greeks were few in number, scattered over a collection of islands in the Aegean and Ionian seas and on the rough terrain of the mainland and the Peloponnesian peninsula. There were also Greek cities in Asia Minor (modern-day Turkey), where important philosophers such as Thales of Miletus (c. 620–c. 546 BCE) and Pythagoras of Samos (fl. 530 BCE) lived and worked. Few of these places had the advantage of a major river and the geographic challenges are part of what forced the Greeks to look outward. Alexander the Great's conquest of the eastern end of the Mediterranean starting in 334 BCE, followed by the capturing of the Arabian peninsula and then on to India, opened trade routes that would establish the Middle East as the crossroads of three continents.

Technology in the Greek world was innovative and also presented complex social problems. Activities that required working with the hands were considered of lower status than intellectual pursuits. This was in part because Greek society depended heavily on slaves who did most of the manual labor and made up a significant portion of the artisans. Because the economy was based on slavery, citizens often had the freedom to pursue specialized interests such as philosophy and invention. Given the social distinction between manual and intellectual activities, inventors could be honored for their intellectual power, but also disparaged for engaging in low-status activities.

Greece produced many important inventors, the names of some lost to us and others such as Theodorus of Samos and Archimedes gaining mythic status. Theodorus of Samos (fl. 530 BCE) has been credited with inventing ore smelting, casting hollow metal forms from molds, a water level, locks with keys and the wood lathe. Since there is no clear surviving evidence of Theodorus' work, and several of the things he is credited with inventing were known (smelting, the water level and the wood lathe) before his time, he may have introduced these inventions to the Greek world, or he may have improved the devices, or he was an actual inventor given credit for more than his own contributions.

There has been a long history of trying to explain why the Greeks became so intellectually, economically and militarily powerful. How could such a dispersed people, living in individual city-states with major internal conflicts and external threats, eventually unify and conquer long-established and larger empires? Part of the answer lies in the organization of Greek civilization into politically independent but culturally related states, and part is a more elusive psychological difference between the Greeks and their neighbors. The city-states, although sometimes at war with each other, were also in a kind of cultural competition for prestige that included intellectual, economic and artistic accomplishments. Greeks were constantly looking for the next great opportunity and they had a history of absorbing ideas from other places and remaking them for their own purposes. For example, the foundations of Greek mathematics and astronomy can be traced to Egypt and Persia, but were transformed into Greek natural philosophy, the precursor to modern science.

The psychological differences between the Greeks and their neighbors also contributed to having a different attitude toward technology. Although the Greeks had a pantheon of gods, these gods were much more human than those of the Egyptians, and crucially most Greeks did not believe that the physical world was imbued with mystical forces. Humans could understand the world and with that understanding they could master the world. The Greeks, especially male Athenians, also lived a very public life centered around the agora or marketplace. The market was not just the place to buy and sell goods, but was also the place to exchange ideas, discuss politics, argue philosophy or look for a physician. Thus, the Greeks were both accustomed to new things and comfortable challenging established norms. While there were limits to dissent, as the death of Socrates for challenging the legitimacy of Athenian government shows, the Greeks generally thought that they could be the best at everything.

The list of Greek inventions is impressive. As a seafaring people, the Greeks improved on or invented many devices related to sailing and navigating, including pulleys, new kinds of ships, the astrolabe, the port crane and the lighthouse. Greek engineers understood gears and produced some of the earliest clocks, watermills and calculating devices such as the odometer and the Antikythera device that was used to follow the position of the sun and moon.

The Greeks traveled all over the Mediterranean world and brought back material goods and intellectual skills such as Egyptian mathematics that they transformed into new work. A large part of this knowledge was held at the Library of Alexandria in Egypt. The library was part of a larger institution known as the Museum of Alexandria. When Alexander the Great conquered Egypt, he established a new capital, and under the ruler Ptolemy I Soter the library was created. It became a hub of intellectual activity at the crossroads of three continents. One of its most famous scholars was Euclid (fl. 300 BCE), who compiled all the mathematical works he could find from Egypt, Babylon and perhaps even India, but his greatest contribution was to organize the ideas into a series of logical steps called "proofs." His writing on mathematics *The Elements* (or *Euclid's Elements*) became the basic textbook for Western mathematics for more than 2,000 years. Eratosthenes (c. 276–c. 195 BCE) would take those mathematical skills and devise a method for measuring the circumference of the Earth that used simple geometry but was very accurate.

And yet, despite this inventiveness, the Greeks are more famous for philosophy and the arts and not as famous for engineering as the Egyptians or the Romans. Part of the reason is that many of the inventions tended to be local developments found in one city-state. The Antikythera device, despite its brilliant demonstration of mechanical engineering, is a single item, not a common device found across the Greek world.

Another part of the reason is that the period of Greek dominance in the Mediterranean world was short. There was not much time for ideas to spread through the unified Greek world before records of inventions were lost, destroyed by war or neglect after the collapse of Greek rule. We can only speculate on what ideas and records were lost forever with the burning of the Library of Alexandria. In addition, the Romans took over many of the Greek inventions and developed their own versions. Despite this, the Western engineering heritage does have a Greek icon in the life of Archimedes.

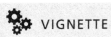 VIGNETTE

Archimedes: The Images of the Inventor

The story of Archimedes (c. 287–c. 212 BCE) is better documented than many other Greek figures, although we don't know much about his early life. Archimedes lived in Syracuse (on the island of Sicily) and was one of the elite of the city, perhaps a relative of King Hiero II. He likely studied at Alexandria, becoming friends with Conon of Samos and Eratosthenes of Cyrene, important scholars at the Great Library in Alexandria. His main interest was mathematics, but he is best remembered for his inventions and the use of displacement of water to solve a problem of density. Included in the list of his inventions are siege weapons including a steam cannon, burning mirrors, the Archimedean screw (a device to lift water), a host of levers, and the block and tackle. He is also famous for working out the principle of the fulcrum for levers.

The problem with the popular stories about Archimedes is that they confuse the historical facts or are simply wrong. The most famous story today is about Archimedes in the bath. The story was told by the Roman engineer and author Vitruvius (c. 80–c. 15 BCE) in Book IX of *The Ten Books on Architecture* (written between 30 and 15 BCE). In brief, the story says Archimedes was asked to determine if

a gold crown made for the king of Syracuse in the shape of laurel leaves contained the amount of gold it was supposed to or whether cheaper material had been substituted. Archimedes, while in a public bath, realized that two crowns with the same volume would have different weights if made of different materials. He leapt from the bath shouting "Eureka" ("I have found it") and ran naked through the city, so intent was he in pursuing his discovery. The crown was proven to be adulterated and the fraud revealed.

The problem is that there is no evidence that Archimedes was ever asked to do this, and no record of the event from his time. It is far more likely that the story was loosely based on Archimedes' research on hydrostatics and buoyancy in his book *On Floating Bodies* (c. 250 BCE). Rather than simple displacement, he could have used a hydrostatic balance, a device he understood, and the mathematics he pioneered. At some level, the reality is more impressive than the story, since a theoretical and detailed knowledge of buoyancy was for more sophisticated science than simple displacement.

Similarly, the story of Archimedes using mirrors to burn Roman ships during the siege of Syracuse has fascinated people for generations. While technically possible, the use of mirrors as weapons seems highly improbable. It would have required the Greeks to have had many large copper or bronze mirrors and there is no evidence of this. Modern tests have demonstrated that mirrors can start fires, but to be effective would have required the Roman ships to sit immobile long enough to catch fire and then have no one notice in time to move or put out the flames. It is also hard to imagine that if the trick worked once, the Romans would have passively let it happen a second time. Historically, there is no mention of burning mirrors from Archimedes' time, so it seems most likely that the idea seemed like something that Archimedes could have done, and was by the process of myth creation turned into something he did do.

It is almost certain that Archimedes did work on a variety of other war machines, including types of catapults. The most famous weapon, although its actual construction is not known, was called the Claw of Archimedes. It was described as a metal grappling hook suspended from a crane arm that could damage ships or perhaps lift them out of the water to tip them over. Some historians have even speculated that the device combined a hook and a trebuchet. Records of the Roman attack on Syracuse during the Second Punic War (214 BCE) describe the loss of a number of ships, but do not provide a clear description of the Greek weapons.

What is true is that Archimedes himself regarded his engineering work as the lowest form of intellectual activity – handwork, not brain work. He owed his engineering skill as a duty to the state and necessary in the defense of Syracuse against the attacking Romans, but it was not the best use of a philosopher's talent. It was his work on philosophy, particularly mathematics, that Archimedes thought was his important work.

We are not certain how Archimedes died, but the most famous story has him working on a mathematical problem when a Roman soldier came upon him. When ordered to surrender, Archimedes was supposed to have refused to move until he had finished his work. In anger, the soldier stabbed him. Again, there is little evidence that this ever happened, but it is part of the myth. What the story of Archimedes' life illustrates is our fascination with clever inventors, especially if they seem somewhat eccentric. In Archimedes' case this fascination was reinforced by the interest of the Romans, particularly Cicero, who in about 75 BCE discovered and restored Archimedes' tomb and celebrated Archimedes as hero.

⚙ A Note on the Lives of Famous Inventors

History is full of mythical tales of inventors, from Archimedes in ancient Greece and Zheng Heng in China during the Han dynasty to Thomas Edison in modern America. There are several characteristics that unite these people. To begin, they all really were inventors, so became known in their own time for actual work. They all faced some form of adversity such as overcoming poverty, facing political rivalry, or living in a time of war. All of them did both practical and theoretical work, so that after their deaths stories could grow up around what the inventor could have done, or might have secretly done, giving them an air of mystery. Finally, over time, other inventions became associated with them that they did not actually invent, such as Archimedes' screw (invented in Egypt), the water clock (originated in Babylon around 2000 BCE, but is often credited to Chinese inventor Zhang Heng [78–139 CE]), or the light bulb (commissioned by Thomas Edison, but not his own work). In the end, the stories of heroic inventors are designed to offer morality lessons for the listeners. The stories tell us about the importance of persistence, determination and being open to sudden insight.

Rome

The defeat of Syracuse in 212 BCE was part of Rome's vast conquests around the Mediterranean basin. By 117 CE, the entire Mediterranean basin was under Roman rule. The empire extended to the Caspian Sea in the east and Britain in the west, north to the Rhine river and south to the Red Sea. When Rome established the larger and much longer-lasting empire than the Greeks, its conquest of Egypt and the Fertile Crescent further strengthened the trade and intellectual ties between Europe, Africa, the Middle East and Asia. The city of Alexandria, already one of the great centers of learning as it combined the ancient knowledge of the Egyptians with the energetic philosophies of the Greeks, was enriched by Mark Antony's gift in 40 BCE of 200,000 scrolls to the Great Library.

Western history tends to give the Greeks great credit for philosophy and the ideals of Athenian democracy, whilst the Romans are remembered for military might, engineering and government. These characteristics are in a large sense true, but a bit more complex since the Roman world subsumed the Greek world and so Greek knowledge (as well as knowledge from other places) helped produce the Roman civilization. We know a great deal about Roman life, because their empire was so vast that there are many physical remains for us to study, from small farming communities to large cities complete with aqueducts and roads, and the monumental architecture of the Coliseum. Natural disaster also helped when the volcano Vesuvius erupted in 79 CE and buried the cities of Pompeii and Herculaneum, preserving them. The Romans also left us many written records that include technological material such as Vitruvius' *De Architectura* and Pliny's *Historia Naturalis*, which, in addition to descriptions of nature, included commentary on mining and aqueducts.

The success of Roman engineering meant that it became a category of employment. People trained to be engineers by being apprenticed to working engineers. Apprenticeship was the most common way to train for almost all professions, including physicians, artisans, cooks and artists. Engineers faced new challenges all the time and had to find solutions for those problems by using trial and error, observation, and the application of basic principles such as a knowledge that water in a container is always level and a plumb bob (a weight at the end of a string) is perfectly vertical.

Facilitated by a growing class of technical workers from surveyors to blacksmiths, Roman industry covered a huge range of devices and systems. In particular, the technology of roads, aqueducts, architecture using the arch, and industrial management contributed to the strength of Rome and its lasting contribution to Western history.

Roads: The Communication Network that Made an Empire Possible

The old saying "All roads lead to Rome" was based on the extensive road system built by the Romans that radiated out from the capital city. The importance of land ownership combined with the large-scale road construction undertaken by the Romans led to the emergence of professional land surveyors known as *gromatici* or *agrimensores*. Both engineers and surveyors used the *groma*, a tool consisting of a rod about 130 cm (4 ft) in height set on the ground and topped with two arms set at right angles making a cross. At the end of each arm was a plumb bob. By aligning the plumb bobs (right and left, front and back) the surveyor knew that the rod was perfectly vertical, and by looking along the arms could map a straight line. The roads were so well engineered that some are still used more than 2,000 years later. There were three grades of construction: paved, gravel and dirt roads:

The classes of roads.

1 *Viae publicae, consulares, praetoriae* or *militares*: main public roads owned by the state, generally connecting major cities. These tended to be paved, such as the Appian Way (*Via Appia*), connecting Rome to Brindisi. Built in 312 BCE, it is named after Appius Claudius Caecus (c. 340–273 BCE).
2 *Viae privatae, rusticae, glareae* or *agrariae*: private roads, although they might be open to the public use for a price. These could be paved, gravel or dirt. Such roads were built everywhere in the empire.
3 *Viae vicinales*: village or district roads. These were often toll roads, but many fell under public control over time. They tended to be gravel or dirt, although they might be paved in towns. Local authorities maintained these roads, so their quality varied depending on the wealth of the communities that controlled them.

The paved roads were marvels of engineering, consisting of a trench filled with different sizes of gravel and capped with dressed (shaped) stone, either rectangular or polygonal in shape. These were laid into the aggregate or cemented into place, the objective being to produce a road surface that needed as little maintenance as possible.

The road was curved from side to side to shed water, and usually flanked by ditches to drain water away. The roads were primarily built by contractors, although the Roman army was pressed into construction duty during military campaigns or when contractors were not available. The main roads, under the control of the state, were maintained by crews based in the cities and towns along the routes. The roads were built with the military in mind, and were often as straight as possible, using engineering such as bridges and embankments to deal with the landscape rather than following the contour of the local terrain. The public could use the main roads, but access was not cheap. Multiple tolls were charged, often at bridges or city gates.

The road system meant that the Roman army could travel great distances and thereby military control of the empire did not require huge numbers of soldiers everywhere. There were garrison buildings across the empire to house the legions and auxiliary forces, for example at Dura-Europos in Syria and Carrawburgh in Britain. If larger military forces were needed, they could be sent out from larger garrisons in or near major cities such as Rome (home of the Praetorian Guard and varying numbers of cohorts) or Alexandria (base for II Traiana legion). Although the ability to move military forces around the empire was important, it overshadows the real power of the roads, which was the flow of information. The Roman road system is a perfect example of what the historian Harold Innis identified as the link between communication and empire. By using a series of mail posts along the main roads where fresh horses were kept, messengers could carry information 800 km (500 miles) in 24 hours. This meant that centralized control was possible, and the government and military leaders had information about everything that was happening across the empire. While military and political information was needed to control the vast empire, the information network also spread knowledge of business opportunities, inventions and discoveries. The roads also contributed to the spread of cultural ideas such as fashion and art, as well as religions, with Isis worship and Christianity moving across the empire by road and ship. The communications network helped keep communities in touch, reinforcing at the periphery the ideals created at the center of the empire.

A good example of the intellectual and practical information that traveled the Roman roads was agriculture. Farming methods and crops were moved across the empire, and the variety of crops was extensive, from grain crops to herbs such as mint and basil, and specialty items such as asparagus and saffron. Pliny the Elder included an entire chapter "The Natural History of Grain" in his *Historia Naturalis*. One of the most important crops exported from Italy to the rest of the Roman Empire was grapes for wine. Grapes had been cultivated for generations around the eastern end of the Mediterranean basin but were imported into western and northern Europe as far away as Britain. All the great wine regions of modern Europe in France, Germany, Portugal and Spain have Roman roots.

✿ Aqueducts

Water is the lifeblood of civilization and controlling water has been one of the most important activities in human history. The oldest recorded aqueduct supplied water to the city of Nineveh in the seventh century BCE, but in the ancient world the Romans were the most accomplished builders. The Romans went to great lengths to supply water to their cities, towns and industrial centers, and the use of aqueducts became one of the most notable aspects of Roman engineering.

The first major aqueduct was the Aqua Appia built in 312 BCE. It ran 16 km (10 miles) from natural springs to the Forum Boarium in Rome. Much of it was underground, carved out of bedrock. Aqua Appia was something of an experiment. It lacked settling tanks or a catch-basin, so it had to be cleaned regularly, and water could not be stored.

The longest aqueduct supplying water to ancient Rome was the Aqua Marcia, completed in 144 BCE. It brought water from 90 km (57 miles) away and used every architectural skill of the era, traveling 80 km (50 miles) underground and then on an elevated section constructed of brick arches for the final 10 km (7 miles). The aqueduct was subsequently repaired and upgraded by a number of emperors, including Augustus and Nero, and in 97 CE it supplied 187,600,000 liters (49,600,00 gallons) of water a day (see Blackman 1978).

Important as the aqueducts constructed to supply water to Rome were, they were relatively small projects compared to some of the other aqueducts built in other parts of the Roman world. The aqueduct at Nemausus (modern Nîmes, France) crossed the Gardon river and required the construction of a triple tier arched bridge 360 m (902 ft) long and 49 m (160 ft) high. The Eifel aqueduct at Claudia Ara Agrippinensium (now Cologne, Germany) linked several water sources and covered 130 km (81 miles). It was almost entirely underground. To build it required Roman engineers to measure the change in elevation over that distance and be able to maintain a near-constant slope in the construction. Two other aqueducts worth noting because they have survived to the modern day are the aqueduct of Segovia, Spain (likely completed by 112 CE) and the Valens aqueduct in Istanbul, Turkey, originally completed in 366 CE and later enlarged several times. Each was built to supply water for a growing urban center, and they remain today as a testament to the skill of Roman engineers and builders.

✿ Architecture

Empires build big, and Rome undertook some of the most massive construction projects in ancient history. They had two technological discoveries that helped them undertake these projects. The first was the use of the arch, and the second was the discovery of concrete. The arch was known before the Romans used it, but it was not

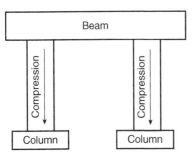

Figure 5.1 Column and beam construction, typical of ancient buildings.

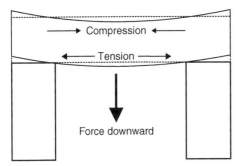

Figure 5.2 Tension and compression forces on a stone beam. Stone beams experience both tension (pulling force) at the bottom of the beam and compression (crushing force) at the top. The longer the beam, the greater the stress difference at the top and bottom of the beam.

much utilized in the architecture of the ancient world in either the European or the Asian world where mass construction was undertaken. The traditional method for building construction that was used by everyone, including the Egyptians and the Greeks, was the post and lintel system, consisting of columns for the vertical elements that supported horizontal beams. The bigger the building, the greater the number and size of the columns and beams needed to support the span of the roof. The entire weight of the roof had to be directed down through the columns. This put the columns under compression, which was good for the stone columns (stone resists compression well), but put the beams under both compression and tension (being pulled in opposite directions) (Figures 5.1 and 5.2). Wood resists tension better than stone, so wood was often used for lintels and roof construction in smaller buildings like houses. The problem for large structures was that wood could be used for the roof, but the weight of the roof was so great that wooden beams to support the roof were not practical, so stone beams were required, which in turn added to the weight on the columns. Massive buildings became a forest of columns, such as the Egyptian temple at Luxor or the Parthenon in Athens.

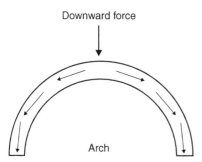

Downward force

Arch

Figure 5.3 Force distribution in an arch. An arch transfers downward force to compression, allowing for a greater distance between columns.

The arch transfers forces from weight above the arch to the ground through columns or piers, and therefore there is little tension and the structure makes the best use of the resistance of the materials (Figure 5.3). This meant that a far greater distance between columns could be achieved. In turn, bigger buildings could be built with the same amount of material as was used in smaller buildings. By stacking arches on top of each other, Roman buildings could also be tall. The Coliseum in Rome consisted of four tiers of arches reaching 48 m (157 ft) high.

Another feature of the arch is that by extending the arch horizontally you can make a vaulted roof. There are several varieties of vaults, such as the groin vault (also known as a cross vault) that is made by intersecting two arches at 90 degrees, such as the vaults found in the Baths of Caracalla in Rome, and the rib vault where three or more arches or half arches are joined and can include arc sections between the arches creating complex patterns, as seen in the Liebfrauenkirche in Mühlacker, Germany.

The final architectural benefit of the arch is the dome. A dome is made by rotating an arch around its vertical axis. Domes were very popular for monumental building because they allow very wide interior spaces without filling the space with columns or walls.

The Romans built with wood, stone and brick, all of which they mass produced, but one of their greatest architectural advances was the use of concrete. Although the Romans were not the first people to use concrete as part of their buildings, they were fortunate to have access to volcanic sands (known as Pozzolana sand) and lime (calcium oxide or calcium hydroxide manufactured from limestone). When mixed in a ratio of one part lime to three parts Pozzolana, according to the Roman architect and writer Vitruvius, the product was a hydraulic-setting cement, meaning that it would harden in water. This made it ideal for foundations, docks, piers and bridges. Cement was mixed with a filler or aggregate such as sand, gravel, broken tiles or even brick rubble to provide volume and strength. The Romans also had sources of pumice (a very light volcanic rock) for making lightweight concrete. After a major fire in Rome in 64

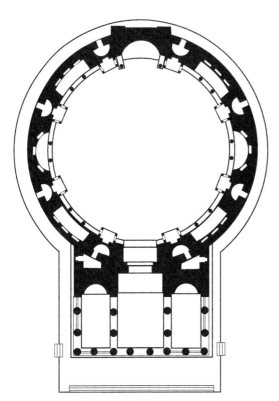

Figure 5.4 Floor plan of the Pantheon, one of the greatest uses of vaulting to create open space.

CE, the Emperor Nero brought in new laws about construction that used brick- or tile-faced concrete construction. These rules led to further innovation in concrete construction such as pouring cement into forms and using wet mixtures for self-leveling floors.

One of the most impressive examples of the dome and the use of concrete that survives from the Roman era is the Pantheon, in Rome, completed in 126 CE (Figures 5.4 and 5.5). The building still exists as the Roman Catholic Church of St. Mary and the Martyrs. The height to the oculus (the circular opening at the top of the dome) is 43.3 m (142 ft), and the dome itself was constructed from lightweight concrete cast in forms. That reduced the weight, but maintained the strength of the dome, and the concrete at the base of the dome used a heavier aggregate of travertine (a form of limestone), and then lighter pumice aggregate for the upper section. At the base, the thickness of the dome is just over 6 m (21 ft), while at the oculus it tapers to just a bit over 1 m (4 ft) thick. The Pantheon was completed in 126 CE, and still impresses visitors to this day, but we are not completely sure who was responsible for its construction. It may have been Emperor Hadrian's architects, but more recent research suggests that it may have been designed by Emperor Trajan's favourite architect, Apollodorus of Damascus, who was the designer of the Arch of Trajan in Benevento, Italy and the bridge of Apollodorus, a massive arch bridge running 1,135 m (3,724 ft) across the Danube.

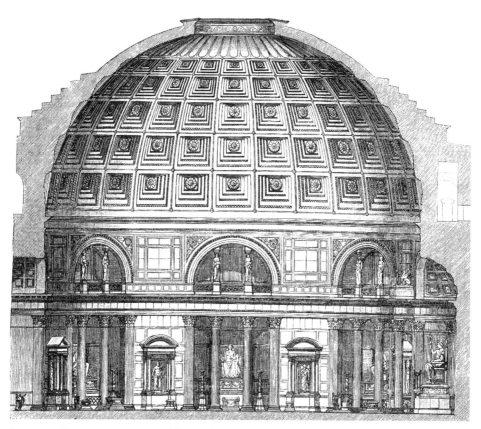

Figure 5.5 Interior cross-section of the Pantheon. The vault used heavier concrete for the thick base and lighter concrete for the upper sections to reduce weight and stress.

✿ Industrial Management

The size of the Roman Empire and the population of the large urban areas required huge quantities of resources, from food and water to cloth and wood. These things had to be transported, often over long distances, so the problem was not just local but spanned the entire empire. This meant that at the heart of the empire, and a driving force for technological development, was the commercial and governmental bureaucracy. As Rome's government changed, particularly from a republic to a monarchy, the leadership changed, but the civil service still had to oversee the empire. They collected the taxes, gathered the information about the status of every city, town and village in the empire, commissioned the building of roads and public buildings, planned everything from baths to entire cities, oversaw the military, and helped keep the people fed. The bureaucracy, unlike what would develop in China, was not fixed or separate from the government. Holding public office was determined by election and patronage. The balance changed under different regimes, with more elected officials during the Roman Republic, which started in 509 BCE and ended in 27 BCE, and more appointments during the empire period. Social status was often determined by the ability to get things done and not just by class at birth, so there was a strong incentive to be productive. The empire consisted of people in a number of classes, with slaves at the bottom, freedmen and citizens, and at the top members of the patrician class (families who claimed to trace their heritage back to the founding of Rome). There was a certain level of flexibility, as those wealthy enough to participate in public life could gain high office in the government and military, and perhaps marry into the patrician class.

A good example of how Romans organized their use of technology to support the empire can be seen at the town of Arelate (modern day Arles, France). Arelate was not a large town, having about 15,000 inhabitants at its height, but it featured a major flour processing complex known at Barbegal. The complex, built on a steep hill, was located north of Arelate where two aqueducts bringing water from the Alpilles mountains joined together. The water drove sixteen waterwheels arranged in two rows, with each wheel above the next down the slope of the hill. The mills were built around 180 CE and continued in use for about 300 years, although some part of the mill works may have survived much later. We have no direct evidence about the amount of grain milled, but historians have estimated that the wheels could produce up to 48 tonnes of flour a day. This was far more than could be used by the town of Arelate, and it was probably shipped around the empire.

Impressive as the mechanical works involved in producing the great amounts of flour were, the site is best understood as a locus of activity that brought together the farmlands of southern Gaul in the Rhône valley, the industrial capacity of Roman engineers, the transportation system including the roads and shipping, and the centralized control system of the Roman government. When Barbegal was rediscovered, it was seen as something of an apex in Roman industry, but in fact there were many such mills across the empire, including flour mills on the same scale outside Rome.

The Roman Army: The Military as a Machine

The vast resources that the Roman Empire used were brought under its control by the power of the Roman army. The military was intimately linked to the culture of Rome, as male citizens were expected to serve in the army, especially in the years of the rise of Rome. The army was made up of standard units from squad level to the legion (Table 5.1).

Table 5.1 Roman army units.

Unit type	Number of men
Contubernium	8–10
Century	80–100 (10 contubernia)
Maniple	120
Cohort	480 (6 centuries)
Legion	6,000 (10 cohorts)

The central military unit of the legion was the heavy infantry that went into battle armed with a spear (*pilum*), short stabbing sword (*gladius*) and shield. The soldiers were highly trained to fight as a unit, using massed forces to break apart enemy units. A typical engagement would see light infantry advance on the enemy to harass and try to break up formations with sling missiles and spears. The heavy infantry would follow, marching on the enemy and throwing their pilums when in range. When the forces came together, the Roman soldiers attempted to cut through the enemy line, killing everyone in front of them and forcing the rest to break up into smaller groups or flee the field. A small Roman cavalry force was used to scout, relay messages and mop up the fleeing enemy.

Military engineers were an important part of the Roman army. They worked on everything from overseeing the building of camps, fording and bridging rivers, undermining fortifications and constructing large-scale weapons. Devices like the ballista, a kind of giant crossbow, and the onager (a catapult that used twisted ropes to provide torsion), that threw large rocks, were often used, especially during sieges. Naval engineering did not come as easily to the Roman military, but by 100 CE, they commanded the Mediterranean with fleets mass produced by Roman shipbuilders.

In exchange for military service the Roman soldier was paid, and when he retired he was given land, at least until all the good land in the empire was occupied. This system displaced many of the conquered people, but it ensured that the military was politically and socially invested in the success of the empire.

⚙ The Fall of the Roman Empire

The fall of the Roman Empire has been closely studied. Many factors contributed to the collapse, although it should be remembered that the empire was divided into a western and an eastern region, and it was the western part that fell first, while the eastern part commanded from Constantinople lasted much longer. Among the factors that contributed to the fall were high taxes, the rejection of military service by Romans leading to the army being made up of people from the conquered lands, political instability and corruption, and the expansion of the empire beyond the limits of information flow. Even the climate may have played a role, as an extended period of cold forced people on the northern steppes to move south, driving in front of them the Germanic tribes that eventually would attack and sack the city of Rome itself.

The technology that Rome depended upon was expensive, both in terms of the cost of making and maintaining the components of the system (waterwheels and mills, roads, ships and so on), but also in terms of the intellectual capital required. Learning how to build the tools and machines took a great deal of time. Technology became a trap. To keep the empire going required that the grand system, particularly the food, water and fuel supplies, keep flowing into the cities. When social chaos disrupted the system, especially the flow of information from the far-flung edges of the empire, the centralized command could not function, and without the political control from the center, the chaos tended to multiply. When the roads were cut off by civil war or by invaders such as the Germanic people who crossed the Rhine in 405 CE, information and resources could not flow in and manufactured goods or the military could not flow out. Cities dependent on aqueducts were vulnerable to attacks that could cut off water to the people. Over a prolonged period, the people who ran the system disappeared, being killed, fleeing to safer territory or simply not being replaced. At first, the machines at Barbegal could be fixed, but with no place to send the flour there was not much reason to repair or even keep using the mill. After a time, there was no reason to learn how to build or maintain big industrial facilities and the skills associated with such engineering faded. It would take hundreds of years before European engineers could match the complex mechanisms of Barbegal or the scale of Roman architecture.

Large urban communities that had depended on supplies from an even larger hinterland could not support themselves, so most shrank. Defense became local as cities and towns tried to protect their immediate locale and were reluctant to send forces to other parts of the empire. Local military forces became more isolated, and independent, looking after their own region. Marginal land became wild as the population declined and the ability to undertake large-scale agriculture was replaced by subsistence farming. The decline of the empire was prolonged, lasting at least 300 years. Many historians mark the end of the western empire as September 4, 476 CE, when the last Roman emperor, Romulus Augustus, was replaced by the Visigoth leader Odoacer.

⚙ The Rise of Islam

Roman influence around the Mediterranean basin, even after the fall of the western empire, helped shape the cultures that followed. In the west, the Middle Ages saw the loss of much of the knowledge of the ancient world but there were both the memory and the physical remains of the great empire, while in the eastern empire the centers of learning lasted longer and never lost as much knowledge. The balance of power that had been under the control of the Romans as the junction of Asia, Africa and Europe shifted dramatically in 629, when Muhammad (c. 570–632), the prophet of Islam, led his forces in the conquest of Mecca. By the time of his death in 632, he had united most of the people of the Arabian Peninsula. The zeal of the Islamic people spread the faith, both by the word as missionaries spread out across the three continents, and by the sword as various caliphs captured territory, eventually reaching from the Iberian Peninsula in the west to India in the east. Although the Islamic world was never completely unified politically, with both dynastic and territorial wars being fought, it still had a high level of cultural unity. The power of the Islamic world was based in part on the religious strength of the faithful, but the empires also grew strong because of trade and the development of a powerful infrastructure that promoted industry, scholarship and the arts. This would lead to what has been called the Islamic Renaissance or the Golden Age, running from about 750 to 1260 CE. The Islamic world had a specific set of rules that contributed to the fast rise and great power gained in this period.

The Five Pillars of Islam form both the religious and cultural foundation of the Islamic world. They provided the Islamic people with cultural and religious continuity, but an unexpected consequence of the Pillars was the creation of powerful intellectual and economic ties across the Islamic world.

1 The *shahadah* or creed that all Muslims repeat to acknowledge the supremacy of God and the Prophet Muhammad.
2 The *salah* or ritual prayer, performed five times a day while facing toward Mecca.
3 *Zakat* or charity, the alms-giving obligatory for all Muslims who can afford it, with the money helping the poor and spread Islam.
4 *Sawm* or fasting during the month of Ramadan.
5 The *Hajj* pilgrimage to Mecca that every Muslim who can afford it must make at least once in their lifetime.

From a religious point of view, the Five Pillars were a brilliant way to unify the faithful and maintain a level of orthodoxy. From a secular point of view, they had equally important effects. Perhaps most important was that they made Arabic the shared language of all Muslims. A great deal of effort was expended to teach people Arabic and the mosques were both church and school. Muslims were encouraged to read the Qur'an. This led to much higher levels of general literacy in the Islamic world than in other places. Although it is difficult to estimate literacy in the ancient world, in the Roman world it was between 5 and 10 percent of the population and it fell to below 5 percent in western Europe after the fall of Rome. We know that by the eleventh century there were bookstores in Baghdad, suggesting that literacy levels were high enough for the commercialization of written material. The interest in literacy also fostered a drive to acquire texts from other places and led to the collecting of Egyptian, Greek, Roman, Indian and Chinese manuscripts by Islamic scholars.

Zakat or charity not only helped finance the spread of the faith and supported the poor, but funded the building of hospitals, libraries, schools and public works. The wealthy had an obligation to the poor, but also to the culture, so Muslim business people and rulers were great patrons, supporting artists and scholars, and often vying to see who could create the greatest public works such as mosques, roads, waterworks, baths and markets. This in turn meant that architects, engineers and armies of builders were in high demand throughout the Islamic world.

The Hajj brought Muslims together from around the world. Some traveled great distances, so they needed a transportation system, and that system, once established, could be used for trade. The pilgrims affirmed their faith, but they also brought knowledge, information and even biological samples from across the Islamic world. They arranged trade deals, encouraged voyages of discovery and mixed cultures. As parts of the old Greco-Roman world came under the control of Islamic forces, Muslims absorbed Greek scholarship and Roman engineering. The skills and knowledge of India and China were also tapped by the spread of Islam. The Hajj helped introduce people to new spices, different foods, tools, fine arts, and crafts. In other words, the Five Pillars helped create fertile ground for the introduction and spread of technology.

❧ The Islamic Agrarian Revolution

One of the greatest innovations of the Golden Age was an agricultural revolution. This saw one of the largest transferals of biological material in history as crop plants and their particular farming needs were transferred from around the Islamic world. A partial list of crops that were transplanted includes bananas, cotton, coconut palms, hard wheat, citrus fruit, plantain, rice, sorghum, watermelons and sugar cane. Many of these crops became big business, such as sugar production that saw dozens of large-scale mills created to crush the sugar cane using presses powered by animals or water, collect the liquid and refine it for local use and export. Cotton grown in Egypt competed with cotton grown in India for export markets as far away as Toledo, Spain.

To support the agriculture, Muslim engineers developed sophisticated irrigation systems using machines such as the chain pump, which worked by drawing discs arranged on a chain through a tube to raise water. To smooth out the operation of such machines, the inventor Ibn Bassal (fl. 1038–75) introduced the flywheel. Perhaps because of the lessons learned in Arabia and North Africa about water use at oases, Muslim farmers avoided the problem of loss of soil due to salt build-up in irrigation systems by a strict regime of crop rotation. Farmers learned about such innovations by reading farming manuals and found out about new crops and their uses by reading botanical encyclopedias such as the *Kitab al-Jami fi al-Adwiya al-Mufrada*. Compiled by Ibn al-Baitar (fl. 1200), it catalogued more than 1,400 different plants and their food or medicinal uses that he discovered or was introduced to during his travels through Spain, North Africa, Turkey and the Arabian Peninsula.

Such manuals and encyclopedias were available in part because paper became a commercial product in the Middle East around 794 when a paper mill started operating in Baghdad. Paper had been produced in China and other parts of Asia for generations, so it was known in the Middle East, but after papermakers were captured at the Battle of Talas (751) that saw Islamic forces defeat a Chinese army in central Asia north of Afghanistan, mills in Samarkand and Baghdad both began to produce paper. The papermakers started adding starch to the pulp, resulting in a paper that was less porous and therefore a better surface for pens than the traditional Chinese paper, which was more suited to brushwork.

Another social factor that contributed to agricultural growth was the freedom of the farmers. Rather than the peasant model found in Europe and China, Islamic farmers primarily operated in a free market. Although it would be incorrect to say that there were no class distinctions in the Muslim world, Islamic philosophy strongly supported the idea of the equality of the faithful. As such, farmers were free to market their goods and there was an incentive to improve agricultural techniques.

With a strong agricultural sector, an increasing number of people were freed to pursue other types of work. Fine crafts, from ceramics to architecture, became the hallmark of Islamic culture, and the techniques were spread throughout the Muslim world. One of the important characteristics of the technological development of the Golden Age was the appearance of the Islamic polymath, a person whose range of activities was very wide and drawing little distinction between practical work and theoretical. This made the Islamic scholars (not all of whom were Muslims) different from the Greeks, for whom there was a much clearer distinction between hand work (craft, and thus lower class) and head work (philosophy, and upper class). An early example of this was the life and work of Jabir ibn Hayyan (c. 721–c. 815). Jabir worked as an apothecary and was responsible for adding a number of plants and animal substances to the list of medical materials. He pioneered distillation, crystallization and filtration for purifying materials, isolating and identifying arsenic, alkali salts, borax, lead, and carbonic, nitric, sulfuric and hydrochloric acids. Many of these were the product of his work on alchemy, but the acids in particular had many industrial uses including leather treatment and paper making. His discovery of aqua regia (hydrochloric and nitric acid) created one of the only liquids that would dissolve gold, making it a valuable tool for the purification of gold and the target of alchemical work for almost a thousand years. His work also contributed to improvements in steel making, and used manganese dioxide to produce clear glass free of the greenish tinge caused by traces of iron found in most glass. Jabir's work was so important and so widely known that he influenced people throughout the Islamic world and in Europe, where he was known as Geber.

In addition to the work that Jabir contributed, his story tells us about a society where the infrastructure had been created to support the kind of concentrated and advanced research that Jabir undertook. At the more mechanical end of innovation, the three Banu Musa brothers (Ahmad, Muhammad and Hasan bin Musa ibn Shakir, c. 803–873) described or invented a range of devices including types of valves, automatons (mechanical figures, often of people or animals), oil lamps, grabs (mechanical jaws for picking things up) and even a gas mask. These were compiled in the *Kitab al-Hiyal* (*Book of Ingenious Devices* or *Book of Tricks*, 850). What is remarkable about the book is not its range or technical sophistication for describing devices that are in common use today, but that the Banu Musa brothers did not regard this work as particularly important, representing a kind of hobby interest in mechanical devices that came second to their main interest in mathematics.

The intellectual backbone of Islamic research was the schools. Mosques functioned as schools that taught language and literacy, and often other skills such as medicine or engineering were associated with certain mosques, but a system of higher education was also created. The first great institution was the Bait al-Hikma or House of Wisdom, founded in Baghdad around 763. Originally a library and center for translation of Greek and Latin texts, it developed into a school, hospital and observatory. By the 820s medical schools began to appear, and in 859 the first madrasah (schools) were founded, starting with Jami'at al-Qarawiyyin (also known as the University of Al-Karaouine) in the city of Fes, Morocco. It is considered to be the oldest degree-granting- university in the world. By 950, these places of advanced education could be found in many major cities.

Not all of Islamic technology originated from the refined research of the schools. One of the most intriguing of the technologies of the Islamic world was the production of specialty steel for weapons. The material, which became known as Damascus steel, produced some of the best blades every created, but the exact process was lost when around 1750 Damascus steel stopped being produced. The power of the finely forged weapons was legendary. There were stories from the Crusade era that Islamic blades could cut through European weapons. While this is highly unlikely, the quality of Damascus steel was superior to anything produced in the West. The weapons themselves were distinctive for having a pattern of lines or waves.

After years of modern metallurgical work, the basic principles of Damascus steel seem to be clear: First, they were made from a particular type of iron ore that had a unique combination of trace elements such as carbon and vanadium; and second, the forging process created an internal structure of fine metal crystals and carbon nanotubes. Despite the name Damascus, the origin of the steel was actually a small number of mines in India and possibly Sri Lanka, and Indian smiths had been making weapons and tools from the ore for generations. The wootz steel (as it came to be called) from these mines was imported to the Middle East and as far away as Toledo on the Iberian Peninsula and forged by a process of heating, folding, hammering and cooling that produced a material that balanced brittle, but hard edge-holding qualities of carbon steel, with the flexibility of low-carbon steel. When the ore ran out around 1750, the process disappeared as well. By that time, the grandeur of the Islamic world had faded to the point that European states were again beginning to challenge the power of the Islamic world and gunpowder had replaced cold steel as the most important weapon.

The reasons for the decline of the Islamic world in the thirteenth century are as complex as those of the fall of Rome, and share some characteristics. Outside pressure starting with the Crusades, but really most importantly the invasion of the Mongols and the sacking of Baghdad in 1258 by Hulagu Khan, disrupted society, ended trade and destroyed resources. Internal political divisions, usually along religious lines, weakened the various states, while the exhaustion of resources in North Africa contributed to a more general collapse. Again, the expensive systems and technology that depended on a wide resource base and the continued training of people to deal with technical and practical problems were disrupted. This was made even worse by the Black Death or bubonic plague that affected large numbers of people in the Middle East. One measure of the damage was that life expectancy among Persian scholars fell from 68 years in 1209 to 57 years by 1242 (Jaques 2006: 187). As times became tougher, the society became more conservative, and when institutions such as universities, libraries and hospitals were destroyed, there was little money or social impetus to replace them.

By the beginning of the fifteenth century, there was another shift in the center of technological innovation. The Middle East, still powerful as the middleman at the crossroads of three continents, was beginning to be circumvented.

1 Why did the Mediterranean basin become a center of innovation?
2 How did Roman roads help keep the vast empire united?
3 What aspects of Islam contributed to the rapid development of technology during the Islamic Renaissance?

✿ FURTHER READING

The rise of empires around the Mediterranean basin has long been associated with the appearance of "new" technologies, even if some of those technologies were invented elsewhere. Broad overviews include John W. Humphrey et al., *Greek and Roman Technology: A Sourcebook* (1998); John G. Landels, *Engineering in the Ancient World* (2000); and K. D. White, *Greek and Roman Technology* (1984). Video sources on Greek and Roman technology are widely available and one of the better collections is *Understanding Greek and Roman Technology* by The Great Courses, 2013. Its twenty-four episodes are lectures with illustrations ranging from naval technology to the building of the Pantheon. After the fall of the western Roman Empire, the rise of the Islamic states absorbed the technology of the earlier states and added new tools and technologies. These are introduced to a western audience by Ahmad Yusuf Hasan in *Islamic Technology: An Illustrated History* (1986).

6 The European Agrarian Revolution and the Proto-Industrial Revolutions

650	Iberian Peninsula under Islamic control
782	Charlemagne asks Alcuin of York to be minister of education
790	Stirrup arrives in Europe from China
910	Abbey at Cluny started
1085	Fall of Toledo
1098	Founding of Cistercian order
1095–1204	Period of Crusades
1347	Black Death begins in Europe
1450	Gutenberg introduces movable type printing
1492	Columbus reaches Americas
1500–1650	Major growth in many European cities
1550	Beginning of Atlantic slave trade
1588	Spanish Armada fails to invade England

Society in western Europe slowly recovered from the fall of the Roman Empire. New technologies were made possible by changes in agriculture that allowed for a growing population and that increased job specialization and consumer demand. The natural catastrophe of the Black Death led to a concentration of money, and that helped finance a growing demand for manufactured goods. The proto-industrial revolution was based on the increased use of waterwheels to grind grain, saw lumber and hammer metals. The Black Death also led to a shortage of scribes, and Johannes Gutenberg created movable type printing to fill the market for written materials. This unleashed a massive change in information technology, increased literacy and sped up the transmission of knowledge. Luxury goods from Asia were expensive but as Europeans became wealthier demand for spices, silk and other trade goods rose. Europeans began to look for ways to trade with China that circumvented the Islamic territories, leading to the voyages of Christopher Columbus. This opened the way for colonization and slavery, which increased the wealth of the European countries that had access to the Atlantic Ocean.

Two revolutions occurred in the late Middle Ages that changed the direction of European society and laid the groundwork for a new age of global empires. In each case, the revolution in fact depended on a literal "revolution." The first case was the agrarian revolution that depended on the rotating of crops, and the second was the proto-industrial revolution that was based on the revolving of waterwheels that offered the first significant application of power that did not depend on muscle power. Although the waterwheel was not new to Europe, as a technology it was revived by the contact between Christian Europe and the Islamic world. Some European farmers alternated field and fallow, but the systematic rotation of crops and the types of crops was also imported. The Europeans built on the technology they copied, modified it, and eventually surpassed the agricultural and industrial capacity of their neighbors. In the early stages of the rise of European power, Asian and Middle Eastern craft and industry were almost completely superior to anything that could be produced in Europe, with two significant exceptions: guns and ships. Those two areas of advanced capability would eventually give the Europeans the ability to take on and defeat the larger and more established powers of the Islamic world and the Asian empires from India to Japan.

In a sense, the use of crop rotation was a matter of observation and utility. In delta regions that underwent periodic flooding there was no need to rotate crops since the soil was renewed naturally. In areas where rice was grown as the staple crop, the use of paddies (rather than dry land farming) with all the water control systems needed pushed farmers toward a different kind of agricultural discipline that led to the introduction of row planting, weeding and the use of terracing to increase the amount of land that could be used for agriculture. In drier regions, constant planting led to over use of land and a decline in soil fertility. Irrigation offered the ability to increase production, but it brought with it the problem of soil degradation over time. In some regions slash and burn agriculture was used, where small plots of land were cleared from forests, particularly rain forests. The problem was that forest land often had thin or nutrient poor soil. Crops were grown until the productivity of the land fell and then the farmers moved on to a new plot after a few years. When populations were small, this system worked as a kind of natural crop rotation, as the forest reclaimed the farmed areas, and over time the cycle could repeat itself. For larger populations, transient farming was insufficient and today such agriculture has contributed to serious deforestation as farmers cut down more forest than can regrow on the abandoned land.

Crop rotation based on letting sections of land lie fallow (not farming it for a growing season) had been practiced as early as 2000 BCE in Egypt, and a form of rotation based on alternating legumes and grain may have been used as early as 6000 BCE in the Middle East. Roman farms alternated crops is some places, but the practice had disappeared in western Europe as agriculture fell back to subsistence levels after the fall of Rome during the fifth century. A more systematic approach to crop rotation had been discovered in Asia and reached a very sophisticated level in the Islamic world, so the idea was likely passed back to Europe through trade or the Crusades.

The form of crop rotation most widely used in Europe during the Middle Ages was a three-field rotation. Farmers planted grain in the fall that could overwinter, such as rye, and then oats or barley in the spring. In the second section, the field was planted with legumes such as peas or lentils, and the third section was left fallow and often used as pasture for livestock.

To understand the rise of Europe to international power, we have to understand the foundation of food production and social change. This takes us back to the fall of the Roman Empire. Under Roman control, farming was a massive and collective activity. Large-scale operations require a competent managerial class or bureaucracy. Such governance structures are a form of invisible technology and large systems cannot survive long without competent administrators. Although Roman farms were largely under private control, production was often highly centralized and the state demanded certain quantities of grain and other foodstuffs from the provinces. These were used to support the military and supply the cities of the empire. The transportation network of roads and shipping was then used to move food around the empire. With the start of the barbarian invasions, the system broke down. After the fall of Rome starting around 450 CE, there was a collapse in the administrative system and a growing isolation of the agricultural lands from the cities. Without sufficient food, cities shrank or simply went out of existence. With no one to maintain and guard the roads, communication and transport became too dangerous or pointless. Without the collective power of the empire, large-scale activities such as building aqueducts and bridges ceased and the engineering skills they required faded. With markets shrinking to just local needs, there was no reason to maintain large industrial facilities such as the mills at Barbegal, and those skills also disappeared. Schools closed, trade waned and skills disappeared for lack of training or use.

As small communities had to fend for themselves, there developed new social structures that combined remnants of the Roman world with indigenous organizations from before the Roman conquests, and elements of the cultures of the invading peoples. One of the foremost examples was the *comitatus*, a system by which a lord ruled with the support of his warriors or knights. The vassals owed military service to the lord, and in return the lord rewarded the vassals with land, money and prestige. The lord also provided services such as legal judgments and support for churches. Although there were hundreds of different local arrangements, collectively they shared enough characteristics that we call the new social system "feudalism." It was one of the most important invisible technologies that shaped European society. Feudalism helped to promote some technologies and limited others.

At the heart of feudalism was the connection between farmers and a local military force. In exchange for protection, farmers supported the local military. In some regions this was done by the military leaders claiming the land and the farmers working it as tenants, often using multigenerational leases for land that stipulated the rent in terms of food production and the amount of work on the land of the landlord. In other places, farmers owned their land, or they could also share land, holding it in common (and hence the term "commons" for public land or public spaces). In that case, taxes were levied to support the military. What started as a means of survival eventually became a hereditary caste system with declining social mobility and rigid rules of responsibility, with peasants at the bottom of the social scale and the land owners (the so-called "landed nobility") at the top, represented primarily by the mounted knights.

The existence of the mounted knight has created an interesting debate about technological determinism. The primary fighting force for the Romans and most of the Germanic tribes that invaded Roman territory was the infantry. Although there had been an elite cavalry force in the Roman army, the cost of keeping horses (and of losing them in battle) made them too expensive to be the main force in the military. The introduction of the stirrup, it was argued by the historian Lynn White, transformed European society. The stirrup gave cavalry a significant advantage in battle since it allowed mounted soldiers to stay in the saddle while using a sword or mace, thus making the mounted knight the supreme military power on the battlefield. According to this theory, feudalism was partly due to the introduction of the stirrup, since many peasants were needed to support the expensive knights, and the knights equipped with the stirrup could more effectively protect the peasants, establishing the network of society obligations. There are some problems with this deterministic argument. The stirrup probably appeared as a way of mounting horses, possibly as early as 500 BCE in India. By 300 CE, stirrups were in use in China and they appeared in Sweden during the sixth century. The stirrup appears to have had only limited effect on military affairs, and it was the case that the Vikings did most of their fighting on foot.

The arrival of the stirrup in central and western Europe came with mounted invaders from central Asia such as the Avars. European forces adopted the new technology to field forces that could fight the enemy on similar terms. By the eighth century, stirrups were in use across Europe. Charlemagne, who beat Avar forces in the 790s, tried to institute a much broader system across the empire he created to support mounted knights, but it did not work very well. Knights on horseback did not completely transform combat until the end of eleventh century. As late as the Battle of Hastings in 1066, the knights rode to battle and dismounted to fight.

The stirrup made the mounted knight a much more powerful military unit, but White's argument that the stirrup was the main agent for the creation of feudalism is too deterministic. There were regions of Europe such as in the Alps and much of the Low Countries that employed few or no mounted knights and were still feudal. It is also the case that knights were often reluctant to fight other knights (who might be relatives, or simply because it was dangerous), so the most effective use of mounted soldiers was against the poorly equipped and almost untrained peasant conscripts who made up most of the armies of the feudal period. This was the case before the stirrup and after, suggesting that the creation of feudalism depended on other factors such as the economics of subsistent farming rather than the transformation of the knights' military power.

While the nobility was consolidating its control through military power, another organization grew out of the era of collapse. This was the Church, or rather the two Churches. In the east, where the Roman Empire did not fall to barbarian invasion, the Church was Greek (and would become what we call today the Greek Orthodox Church), while in western Europe the Church was Latin (the foundation of the Roman Catholic Church). The Church ministered to the spiritual needs of everyone, was the source of medical care, and because it maintained literacy, provided record keepers, teachers and scribes.

The life of the early Middle Ages was often dire. Invaders pressed from the north and east, and after 650 CE from the south and west as well, as the Iberian Peninsula came under the control of Islamic forces. Farming at just above the subsistence level meant that there was little surplus and thus little support of a culture of artisans or craftspeople. What surplus there was went to support the military class, particularly the mounted knights. Psychologically, many of the people of western Europe believed that they were in the end days as foretold by the Bible, and there was direct evidence that the past was better than the present. Bridges, roads, aqueducts and buildings, even if in ruins, were beyond the ability of the feudal system to replicate.

One of the great turning points in Western history was the rise of Charlemagne. Charlemagne came to power in 768 CE and his goal was to recreate the Roman Empire. In addition to being a powerful general, he was a good politician who understood that conquest was not enough to build an empire. He worked to establish courts of law and government ministries to oversee his empire, creating an infrastructure for stability and confidence. One of the common problems across western Europe was the low quality of many priests or the complete lack of priests in some regions. In 782, Charlemagne asked Alcuin of York, one of the most learned men of the age, to be his minister of education and organize a school system. Alcuin built a system that placed the responsibility for educating priests on the shoulders of the bishops by requiring them to establish cathedral schools.[1] In turn, local schools were set up in larger churches and in the growing number of monasteries that were established during the relative peace of Charlemagne's rule. The church schools taught basic Latin literacy, and sent the brightest boys on to the cathedral schools to be more fully trained, either becoming priests, joining monastic orders, or in some cases finding work at court. This school system (another invisible technology) created the training ground for European public servants.

The church functioned a bit like a transnational company, operating a mail system, moving information and innovations across Europe, keeping records, and offering scribal services and education to the nobles and government officials (and thus keeping track of secular events). The monasteries became centers for technological preservation and development, saving Greek and Roman texts, and many of the skills such as stone working and the secrets of waterwheels. Monks introduced Merino sheep, bred cattle, pigs and dogs, began terrace growing of grape vines on marginal land and then produced wine in large quantities using mechanical presses.

Some of these innovations were original discoveries, some were recovered knowledge that came from the records and documents that the Church had saved from the Roman era, and some were acquired from the Islamic world and Asia. The exchange between the Islamic and Christian areas often seems strange when the two religions were so often violent rivals, but two things intervened. The first was that Islamic scholars, when political relations permitted, allowed European scholars to visit and study at centers of learning such as Toledo and Jerusalem. They believed that even though Islam was the true religion, Muslims, Christians and Jews were the People of the Book, who shared a common bond through the documents of the Old Testament. In addition to this scholarly contact, trade was conducted between Europe and the Middle East, particularly through the Byzantine Empire and by the Venetians. Gold and silver in the long run transcended most military conflicts.

Charlemagne's empire was short-lived, as his sons fought among themselves for control after his death, but the psychological and educational impact of his rule was much longer lasting. After Charlemagne, there was a much stronger sense that Europe could achieve great things. This started to be reflected in the size of building projects, such as the Benedictine abbey at Cluny, France founded in 910 CE. The abbey would grow to be one of the biggest construction projects of the era, and become a center for learning and influence until the thirteenth century.

Another turning point in European history occurred in 1085, when Alfonso VI of Castile captured the city of Toledo, the capital city of the Taifa kingdom of al-Andalus. Under Muslim rule, the city had become one of the leading centers of learning in the Islamic world. Fortunately, the libraries at Toledo were not destroyed by the conquerors, and their capture sparked a major period of intellectual growth in western Europe. Although it would be several generations before all the Muslim territory was captured by Christian forces on the Iberian Peninsula, the victory at Toledo gave European leaders a sense that they could challenge the strength of the Islamic world.

In addition to the military impetus, there were wider reasons that Europeans began to look out at the rest of the world, starting at the end of the eleventh century. Most important, the pressure of invasions had ended, and there was relative peace in Europe. That meant there were lots of knights with not much to do. The agricultural situation was good, with a significant portion of the best land in production, providing the surpluses needed to support the knights and a growing number of workers. The downside of the agricultural situation was that most of the good land was in use, and by the rules of primogeniture first sons inherited the land, while subsequent sons had to work for a living, there being no good land left in Europe to develop or capture. To be a member of the landed nobility but hold no land was a problem, so capturing territory in the Holy Land was a significant consideration. Trade was on the rise, so action to secure trade routes and monopolies was an inducement to fund overseas adventures. And there was the traditional enmity between Christians and Muslims, made worse when the Seljuk Turks attacked Christian pilgrims to Jerusalem.

These factors came together in 1095 with the First Crusade (1095–99). Christian forces captured a number of coastal cities and took Jerusalem in 1099, sacking the city and massacring the inhabitants. The Second Crusade (1147–49) failed to take any territory in the Middle East, but forces did capture Lisbon in 1147. The Third Crusade (1187–92) started when the legendary leader Saladin recaptured Jerusalem, but it led to only a few minor victories for the Europeans and left Jerusalem in Muslim hands. The later Crusades were even less successful, including the sacking of Constantinople (a Christian city) by a Crusader force in 1204 that confirmed the Great Schism between the Latin and Greek branches of European Christianity.

What the Crusades did from a technological point of view was expose Europeans to the wonders of high technology. Table 6.1 lists some of things the Europeans encountered because of the Crusades.

The agricultural revolution in Europe was heavily influenced by Islamic practices, and in particular by irrigation techniques, both for watering land and for draining it, which helped transform Europe's subsistence agriculture into a much more productive industry.

Table 6.1 Technological acquisitions from the Crusades.

Chess

Astronomical instruments including the astrolabe, sextant and quadrant

Silk

Distilled alcohol

Gunpowder

Soap and perfume

New crops including sugar cane, rice, cotton, artichokes, egg plants, citrus fruit and almonds

Weight-driven mechanical clocks

Hospitals and surgical instruments

Lateen sail (triangular sail hoisted on a vertical mast)

Mills using waterwheels and wind power, including saw mills, gristmills and paper mills

Glassmaking

Paper

Spices including pepper, cinnamon, cardamom and cloves

Europeans made use of a number of tools they found in Islamic Iberia, such as various types of noria. The noria was a wheel with pots or buckets arranged on its circumference that lifted water from a river or pond and dumped out the water into an irrigation channel or aqueduct. It could be driven by animals, or by wind, or by water power using an undershot waterwheel to lift the buckets.

The Cistercian Order: Industry and Religion

The technology of the waterwheel was spread across western Europe by way of religious orders such as the Benedictines and the Cistercians. The Benedictine order was founded about 530 CE and followed the Rule of St. Benedict that set out the tenets of monastic life. The Benedictines grew rich over the years, as they were given land, often in wills, and turned that land into productive farms and other businesses. They also made money catering to the pilgrimage trade. This allowed the Benedictines to take on big projects such as the building of Cluny Abbey. As a reaction to what some saw as too rich a life for the true believer, in 1098 the Benedictine abbot Robert of Molesme and some followers established a new monastery at Cîteaux (Cistercium in Latin), south of Dijon, in France. The order led a very simple life that included manual labor, but in part because of their proximity to Iberia, and information coming back from the Crusades, the Cistercians adopted the most modern of agricultural ideas and industrial practices. They became renowned farmers, introducing new crops and farming methods (including crop rotation), and breeding horses and cattle. Meat production was purely for profit, since the order was largely vegetarian. Each monastery was a tiny agricultural and industrial hub, where in addition to the abbey, there was a production house powered by waterwheel. These wheels were used to grind wheat, to cut wood, to full cloth (beating it to increase its density), and later to pulp linen rags for paper making.

A good example of the Cistercian approach was the Real Monasterio de Nuestra Señora de Rueda (Royal Monastery of Our Lady of the Wheel) in the Aragon region of Spain. Founded in 1202, it had farm land, a salt works, a vegetable oil mill, a flour mill, a vineyard and wine cellar, and orchards. Production was powered by a large waterwheel using water diverted by a dam on the Ebro River.

With such a successful system, the Cistercian order experienced explosive growth. By 1152 there were 333 Cistercian abbeys across western Europe. There was so much growth that in that year a halt in the founding of new abbeys was enacted by papal command, but the pressure for growth was too much and by 1200 there were 525 Cistercian houses. The Cistercians ran what amounted to a vast international business, even operating a fleet of ships to move people and produce around Europe. In Flanders, one of the great centers of medieval commerce, the Cistercians farmed more than 10,000 hectares (about 25,000 acres).

Keeping track of all the developments and making sure that the members of the order were following the rules was a major managerial problem. The Cistercians needed to keep written records, and so they trained and employed monks as scribes and accountants. The scriptoriums of the churches and monasteries were constantly busy copying the Bible and other documents. Like the Chinese scribes, they were an elite class, but because of their religious rather than secular nature they were less integrated into the secular bureaucracy. Over time, a growing number of independent schools became the first universities. Oxford, Paris and Padua were the earliest schools to gain formal standing as universities, going on to teach generations of priests, scholars and administrators.

✿ The Black Death

Western Europe had slowly recovered from the collapse of the Roman Empire, but the period of stability was suddenly interrupted. After years of good weather, in the 1340s global temperatures fell, and Europe suffered from several years of cold, rainy summers that devastated harvests. This was made worse by the spread of wheat rust, a fungus that killed wheat crops or poisoned the grain. In their weakened condition, Europeans were ill-prepared for the Black Death. This was an outbreak of bubonic plague that arrived in Europe in 1347, most likely carried from Asia onboard trade ships crossing the Black Sea. It reached Italy in 1348 and spread to the rest of the Mediterranean basin in the next year. The disease was carried by fleas that lived on rats, and when the rats died, the fleas found human hosts. The speed of the disease was incredible, with infection leading to death in as little as 24 hours.

When the Black Death struck Europe and the Middle East, arriving in Alexandria in 1347, it devastated many cities, where most of the merchants, artisans and scribes lived and worked. More than a quarter of the population had died by 1351 when the plague receded. No one really knows what the death toll was, but estimates run from 75 to 200 million people in Eurasia. As the historian James Burke pointed out, terrible as the Black Death had been, when it ended everyone was ready to celebrate and forget the horrors. And they had the money to do so, since the dead had left their property to the survivors. Clothes were purchased, expensive spices were imported, and land was redistributed. Because of the increasing trade with the Middle East, there was also a growing interest in finding a way to avoid the middlemen of the Venetians and Arabs on the great trade routes to Asia.

The end of the plague also brought together three things that set the stage for a powerful new technology to be created. The three things were cheap linen rags from all the clothes and underwear everyone was wearing, the spread of the tools to make screw threads used in everything from wine presses to clocks, and a shortage of scribes.

❀ Movable Type Printing

Scribes had lived mostly in urban areas and were hard hit by the Plague, so just at a point when the demand for scribal work was on the rise, the number of scribes was very low, driving up the price of employing them. This created an opportunity that changed European history, when Johann Gutenberg (c. 1398–1468) created the first successful movable type printing press. There is some speculation that Gutenberg knew of the press from other sources, since printing was used in China, Korea and the Middle East, but his work seems to have been based on figuring out how to combine a number of other inventions rather than a knowledge of some working press. The various techniques, such as paper making, block printing and the screw press were all inventions that had filtered out of China, but putting them together was only part of the solution. It is the case that Gutenberg was a silver smith and had a special set of skills that made it possible for him to construct the individual letters, the key component of the system. As a smith, Gutenberg knew about the production of stamped coins and the hallmark, a symbol or series of symbols stamped into fine metal work to identify the maker. This was done with a carved punch. He also knew about casting fine metal. Putting the two ideas together, he created a set of punches with the letters of the alphabet in lower and upper case. He punched the letters into softer metal (brass) and then put them in a mold, casting as many of each letter as he needed from white metal, an alloy of lead and tin. These were set in a frame to hold them in place. With the addition of the numbers and some punctuation, Gutenberg could replicate anything the scribes could write. Unlike the large print sets that the Chinese needed, Gutenberg needed only about sixty characters.

Having created movable type, he needed two other things: ink and a press. He experimented with many ink formulas, but the exact content was one of his greatest secrets. Water-based ink would not adhere to the metal, so Gutenberg's ink was oil-based. He may have borrowed techniques from artists, who in this period were experimenting with pigment formulas for oil painting. The ink needed to be thick enough to stick to the characters and not pool in the carved-away parts of the letters or bleed on the paper (spread out in a blur because of the capillary action of the paper fibers), but it had to be liquid enough to transfer completely to the paper so it would not clog up the print elements. Gutenberg's ink was primarily made of carbon, oil (linseed or walnut being the most likely) and traces of lead and copper. There may also have been traces of turpentine and lead monoxide.

The press was based on the screw system that had originated in China and been used by European monks (particularly the Cistercians) to produce wine. Gutenberg took the screw design, and instead of spiraling down to press the grapes it traveled a short distance to press the paper. The type was set in trays and laid flat on a platform. The ink was applied, the paper was placed on top of the type, and then a pad, likely made of leather over a wooden or metal plate, was pressed down using the screw system. The press was released and the paper pulled out and hung up to dry.

Gutenberg noted that one of the simplest and most repetitive jobs being done by scribes was producing indulgences. These were small pieces of paper given to people – for a small donation – as a token of the remittance of minor sins. Selling indulgences made the Church a great deal of money and would be one of the things that triggered Martin Luther's objections to the way the Church was run. Gutenberg was printing such material by 1449, but he was not satisfied with the lowly job of bulk indulgence production, so he turned to printing psalms and then undertook to print bibles. The Gutenberg Bible (also known as the 42-line Bible or the Mainz Bible) was printed in 1454 and sold by subscription, but the cost was too high and Gutenberg eventually went bankrupt. He was eventually forced to give his press to his creditors, and Johann Fust (c. 1400–66) took over his business. Except for some legal documents, that is the last thing we know about Gutenberg, who died in obscurity and likely in poverty.

The printing press tapped into a large and growing demand for books and other printed materials. This could have made Fust and Gutenberg's other original backers fabulously wealthy, but they were unable to control the spread of the invention. In modern terms, almost every part of the printing press was "off the shelf" technology. Anyone who had any mechanical ability who saw a printing press could replicate it, and the idea and the components were quickly copied. By 1474, William Caxton was printing books in England. By 1500, only fifty years after Gutenberg printed his first lines of text, 5 million books had been printed in Europe.

One of the most innovative printers was Teobaldo Mannucci, better known by his Latinized name Aldus Manutius (1449–1515), founder of the Aldine Press in Venice. He introduced the italic font to create a more compact print form, and created the octavo (around 13 × 20 cm or 10 × 15 cm; 5 × 8 in or 4 × 6 in) size book, the precursor of the modern pocketbook, that was small enough to fit in a jacket pocket or horse saddlebag. He was also important for publishing many of the Greek and Roman classics, leading in part to the revival of interest in the scholarship and ideas of the past, including a renewed interest in architecture and engineering marvels. This was part of the larger social and cultural transformation of the Renaissance.

As a technology, printing reinforced and helped spread other forms of technology such as education, engineering, accounting, architecture and medicine, to name just a few. It also helped promote scientific ideas by first introducing a much larger audience to natural philosophy from the Greeks, and then giving a platform for those who wanted to challenge the old ideas. René Descartes and Galileo would not have had such an impact on European thought if their ideas had not been printed and quickly distributed.

In addition to what was printed, the act of printing by itself changed European life. It was an information revolution, taking the control over the written word out of the hands of the clerics and scribes. Although the Church attempted to control printers by creating a system of censors to review books, it had trouble controlling the widespread and relatively mobile technology. In 1559, the Church established the Index of Proscribed Books. This listed the books that the faithful were not supposed to read without special permission, but it did not take long for there to be too many books produced in too many places to control. As the price of books fell, the levels of literacy rose. Languages, especially the vernacular languages (languages spoken by the common people in a region), became more important, and Latin, which had been the language of scholarship and the Church, became less important. Printing codified language by creating educational material such as primers and grammars, and by making spelling consistent. Although printing spread vernacular languages, it also squeezed out regional differences, so that, for example, Paris printers dominated the market, making the French of the royal court and Paris became the French of all France.

Printing also changed people's relationship with information. Books were more accurate than hand-copied materials, and if there were errors it was much simpler to change type than to fix a manuscript. As printed material grew longer, printers needed some way to make sure that the pages were in the right order. This was because the print sections were divided up into subsections (called "quires") and the order of the pages was complex as more than one page was printed on a sheet of paper, and print was applied to both sides of the sheet.[2] This led to the introduction of the page number. Printers, always looking to find the most efficient system and use the smallest number of characters, began to use Hindu-Arabic numerals rather than Roman numerals. The Hindu-Arabic numerals had been introduced to Europe around 780 CE, but did not gain wide use until after printing made them popular. The addition of page numbers meant that indexes and tables of contents could be created. A person no longer had to read an entire scroll or an entire manuscript to find out a particular piece of information. A reader could use the information-locating power of indexes and page numbers to look up the desired information. This had the effect of separating information from the context of the larger work.

The great flowering of art, architecture and literature of the European Renaissance was driven by a rising economy and the print revolution. While we marvel at the brilliant arts, the Renaissance was also a time of great tension. External threats pressed in from the east and north, with the Mongols controlling Muscovy, and Islamic forces on the eastern boundary through the Balkans that would reach the gates of Vienna in 1529 before falling back. The Mediterranean was a dangerous place and the trade routes to Asia were making the empires in the Middle East very wealthy. Internally, the Reformation and the Counter-Reformation created a religious cold war that split families and set nations against each other.

✿ Crossing the Atlantic

To the west, the Atlantic lay open, offering access to Africa and the tantalizing possibility of a route to Asia that avoided the Venetian middlemen, pirates in the Mediterranean and the Islamic world. It had long been known that something existed across the ocean, since the Vikings had traveled in the north, and fishing boats from England and the Basque region had reached the Grand Banks. The problem was that such stories lacked detail and were mixed with fantastic tales. Yet the lure of adventure and enormous wealth proved too attractive to ignore. All it took was a bit of ingenuity and ships strong enough to brave the Atlantic.

When Christopher Columbus planned to sail for Asia in 1492, he had a copy of the tales of Marco Polo, but on more practical grounds, he read Ptolemy's astronomical work the *Almagest*, and understood stellar navigation. The *Almagest* had been known from the Middle Ages, but Columbus had also read *Geographia*, Ptolemy's treatise on cartography and atlas of the world as known by the Romans. It had only recently been rediscovered and printed. Putting the theoretical and technical ideas together with the lore of the sea, Columbus reasoned that he could reach Cathay, which he calculated was only 1,100 leagues (about 4,727 km or 3,000 miles) west of the Canary Islands.[3] Since sailors and scholars understood that the world was a sphere, this was not a ridiculous idea. The big argument that Columbus had with the scholars of the day was about the size of the sphere. Columbus thought that the Earth was small and he could reach Japan in a matter of weeks. A panel of scholars at Salamanca, Spain argued that the globe was much larger and it would take a very long time to reach Asia by going west, advising the Spanish government to reject Columbus' plan.

As it turned out, the scholars were right and Columbus was lucky that there was a continent to run into or he might never have returned. His 1492 voyage did not result in the western route to Asia, but it opened the way to the Americas. In 1493, Pope Alexander VI drew a line on a map that gave all undiscovered land to the west of the line to the Spanish, while to the east the new lands belonged to Portugal. The fact that the lands were held by people who had their own civilizations did not deter the European adventurers. In fact, the discovery of cities in the New World spurred on the Conquistadors who, aided by the technology of gunpowder weapons and the swift spread of diseases to which the peoples of the Americas had no resistance, looted the cities and enslaved the people.

The Spanish could consider such an adventure in part because they had access to a new class of ship, the caravel, that had been developed by the Portuguese. The caravel was usually a two- or three-masted ship with triangular sails set in lateen fashion (following the rigging style of the Egyptian dhow), but caravels were also built with square rigging. An older ship type, the carrack, which was a three- or four-masted ship with square-rigged sails, had been used in the Atlantic, but usually as coasters, rarely sailing out of sight of land. Carracks were not as maneuverable as caravels, and so were less useful for exploring. The two ship types were both designed to withstand the waves of the open ocean, having higher sides and prow, and carrying more supplies than galley-type ships used on the Mediterranean. When the routes to the Americas were established, it became more economical to build larger and larger ships. The carrack and caravel evolved into the galleon, reaching 2,000 tonnes displacement and often purpose built as either a warship or a transport. The larger size actually made the galleon faster than smaller ships (more sail and easier motion through the water), and the introduction of the stern-mounted rudder made it much more maneuverable (Figure 6.1).

The pintle-and-gudgeon rudder was one of the inventions that allowed the European sailors to undertake their voyages. The rudder had evolved from steering oars that had appeared around 3100 BCE in Egypt. Quarter rudders were large oars used over the side of the boat, that were later moved to the stern. Stern rudders, rather than specialized oars, were probably first invented in China sometime around 100 CE. They were not attached to the stern by a hinge, but sat in pockets or sockets, and were suspended from above the waterline. The European rudder consisted of a long, flat section of wood

Figure 6.1 Carracks (top left), a galleon (top right), galleys (bottom left) and a fusta (bottom right). European naval technology changed to meet the conditions of sailing the Atlantic Ocean.

attached vertically to the stern of the ship by a long pin going through iron hinges and operated by ropes, and, as ships got larger, block and tackle. These control ropes were later connected to a steering wheel that used gears and capstans to move the rudder against the weight of water. In addition to making ships more maneuverable, the rudder helped sailors keep the ship's bow into the waves, a necessary part of open ocean sailing.

The scale of even the smaller ships meant that governments and investors had to commit significant resources to enter the maritime game, and, for businesses, building a galleon was such a big investment that few merchants could afford to be sole owners. Ship building thus also contributed to the creation of corporations, as groups of merchants pooled their money and spread their risk to build ships, and then worked out the principles of shares to distribute the profits.

In doing so, the Spanish changed the economy of Europe. Over time, gold and silver had been flowing out of Europe to the Middle East and Asia. The Europeans had few trade goods that the people of the Chinese, Indian or Islamic empires wanted, so the luxury trade in spices, silk, porcelain and other goods was almost entirely based on gold and silver. Europe sent tons of gold and silver out to the East to pay for these goods. Without bullion for coins, it was not really possible to have a cash-based society except in a few places such as Venice. Most trade in Europe prior to 1492 was barter and trade in kind, and farmers paid their rent with labor and food. With the conquest of the Americas, hundreds of tons of gold and silver began pouring into Europe. In current dollars, Spain had gained the equivalent of $1.8 trillion from its conquests in the Americas by 1600. This led to a far greater use of coinage, leading to a more general form of capitalism, and a greater ability to finance industry by investing. The Conquest started the Atlantic era, when Europe turned its focus away from the Mediterranean and Asia. The organization of business interests and the cash would help finance the Industrial Revolution.

It is difficult to overestimate the importance of the transformation of the European economy that came about because of the discovery of the "New World." The influx of bullion financed big projects and changed social relations. The landed upper class paid for their new lifestyles with gold and silver and in turn wanted cash rather than crops and labor from their tenants. Without labor and produce as the basis for renting land, peasants were not tied to the land to the same degree. Many laws were passed to force peasants to stay where they were, but over time these all failed, as people began to move for work, to find better economic circumstances or, in the case of skilled workers such as weavers and glass makers, were invited by rich patrons to relocate.

The Spanish experienced massive inflation in what economists have called the "price revolution." In the 150 years following the voyages of Columbus, goods increased in cost by an average of 500 percent. This affected all of Europe as the Spanish purchased goods from other places and spent large sums of money fighting wars. Bullion flowed out of Spain, and colonization began to create a new market for European products as the Atlantic began to replace the Mediterranean as the main focus of international trade.

One of the events made possible by Spanish wealth was the plan to invade England. In 1588, the Spanish built and sent an Armada of 130 ships to Flanders to pick up an army for the invasion. The invasion failed, from a combination of poor planning, lack of communication, bad weather and superior English naval tactics and weapons. In particular, many of the English ships were faster, were armed with better cannons and, crucially, could load and fire their cannons during battle. As the military historian Geoffrey Parker noted, the Spanish loaded their cannons ahead of battle and planned for one crippling salvo followed by ramming and boarding the enemy ships, but the

English bombarded the Spanish fleet from a long distance (see Martin and Parker 1999). One should be careful about attributing too much of the victory to technology, however. In 1589, Elizabeth I sent an English fleet against the Spanish with the hope of igniting a revolt against Spain by the Portuguese. The attack failed, despite the advantage of better ships. Both countries were going into bankruptcy from the cost of the war when they signed the Treaty of London in 1604.

The "Putting-Out" System and the Origins of the Proto-Industrial Revolution

The influx of bullion contributed to what has been called the "proto-industrial revolution." It was much easier to finance businesses with cash at just the moment that the price of goods was on the rise because of the price revolution. One part of the revolution was the increasing availability of water power technology to run a greater and greater variety of mills and businesses. Everything from saw mills to tanneries used water power to process materials. This required the builders to overcome the problem of gears and cams. Gears allowed the speeding up or slowing down of the rotation of the waterwheel to suit the work and compensate for changes in the rate of flow of the water. The cam, at its simplest just a peg in a rotating wheel, allowed the engineers to turn circular motion into linear motion, and to create reciprocating action such as hammers and pulpers.

Merchants could more easily conduct business in a cash-based economy, and during the proto-industrial revolution there began the "putting-out" system (also called the workshop system or the domestic system). This was particularly important for the textile industry, but was also used for many other products, including shoes, hand tools, buttons and firearms. In the putting-out system, a merchant would supply raw materials such as wool to families who would spin, weave and finish cloth in their homes. This was a welcome source of income, especially in the winter when farm work was not possible. The merchant then picked up the finished goods and took them to market. The workers made cash to supplement their agricultural income, while the merchant had the flexibility to scale production to markets, increasing or decreasing production as needed.

In larger terms, the proto-industrial revolution represented a slow trend toward industrial labor. To increase efficiency, there was an increase in the division of labor as parts of manufacturing were divided by workshop. For example, a merchant would have wool cleaned and carded (combed to straighten the fibers) by one family and spun into yarn by another workshop. At the same time, the expanded use of water power demonstrated the utility of non-muscle power for work. It also allowed producers and governments to consider activities that would require large-scale production of goods, whether it was naval vessels or cooking pots. In turn, increased production led to an increasing demand for natural resources. Some of these demands were met by more intensive domestic exploitation of agriculture, mineral resources and forestry, but it also put more pressure on governments and merchants to find resources from other places. With ships built to withstand the conditions of the Atlantic, cod was fished on the Grand Banks, wood was imported from Scandinavia, sugar from the Caribbean and furs from North America, to name only a few imports.

✿ Slavery

One of the tragedies of the proto-industrial era was the emergence of the slave trade as a major component of industrial development. Sometimes called the "triangle trade," it saw the newly financed overseas traders taking manufactured goods from Europe

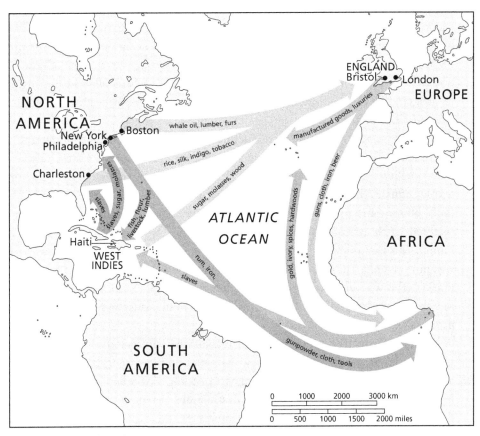

Map 6.1 The triangle trade.

Table 6.2 Estimated slave imports into the Americas by importing region, 1519–1866 (Rawley and Behrendt 2005: 368).

Destination	Slaves
Brazil	3,902,000
British Caribbean	2,238,200
Spanish America	1,267,800
French Caribbean	1,092,600
Guianas	403,700
British North America	361,100
Dutch Caribbean	129,700
Danish Caribbean	73,100
Total	9,468,200

to Africa and exchanging them for slaves. In turn, the slaves were imported to the Caribbean and the Americas as workers in agriculture and mines. A variety of products such as sugar, rum and later cotton were then shipped back to Europe. Note that in the triangle trade bullion does not flow out of Europe, but at the same time the system encourages manufacturing in Europe.

Slavery was a technology. Although there have been slaves through much of human history, the Atlantic slave trade was created to solve the problem of finding low-cost labor to work in the harsh conditions of the Caribbean and the Americas. The system was so effective because European trade goods, particularly firearms, were highly valued and had the power to change social relationships in Africa, as groups that gained access to European technology and products were able to dominate their neighbors. The slave trade left large sections of West Africa depopulated and in constant turmoil, which in turn made it harder to resist enslavement.

Table 6.2 represents only those slaves who made it to the destination. Although the total number of people enslaved will never be known because there are almost no records outside those that the slave traders kept, it is likely that between 1.2 million and 5 million people died while enslaved but before reaching the destination.

The profits of the slave trade helped to finance the technological revolution in Europe. In particular, the sugar trade generated huge wealth for the plantation owners and sugar traders. For a brief moment, Hispaniola (modern-day Haiti and the Dominican Republic) was one of the most valuable places on Earth.

❧ Population Change in the Proto-Industrial Period

One of the ways to see the changing place of power in the proto-industrial era is to look at the growth of European cities (Table 6.3). From the end of the sixteenth century, the major population centers in Europe shifted from the Mediterranean region, particularly Italy, to those associated with the Atlantic trade.

Cities, from an economic point of view, drive consumption. The population needs goods and services, and as they grow, the hinterland needed to support them must either grow or produce more. During the proto-industrial revolution, the hinterlands did both. Long-distance trade in the Atlantic increased, bringing in resources and spurring economic change. Closer to home, food production increased because of better farming methods, and manufacturing grew because of the increased use of water power by mill operators and a more efficient network of producers.

Table 6.3 Population of European cities, 1500–1650 (based on Chandler, Fox and Winsborough 1974: 83–299).

City	1500	1600	1650
Amsterdam	25,000	48,000	165,000
Lisbon	55,000	55,000	170,000
London	50,000	187,000	410,000
Milan	89,000	107,000	105,000
Naples	114,000	224,000	265,000
Paris	185,000	245,000	455,000
Seville	70,000	126,000	60,000
Venice	115,000	151,000	134,000

1 Why was food production so important to the increase in innovation in Europe?
2 How did the Black Death contribute to changing the place of technology in society?
3 What role did the slave trade play in the industrialization of Europe?

❧ NOTES

1. Although we often associate cathedrals with large, beautiful churches, a cathedral is an administrative designation meaning the home church of a bishop.
2. For example, to print an eight-page booklet on two sheets of paper with four pages printed on each sheet (two on each side of the paper), the pages are printed 8, 1, 2, 7, and 6, 3, 4, 5.

3. If Columbus could have sailed straight west from Spain to China, he would have covered 8,800 km (5,474 miles).

FURTHER READING

The Middle Ages in Europe has been one of the primary focuses for European scholars trying to figure out the relationship between technology and the structure of the modern world. Karl Marx argued that feudalism was (at least in part) a product of the demands of the tools used at the time. Frances and Joseph Gies produced a popular book looking at medieval technology, *Cathedral, Forge, and Waterwheel: Technology and Invention in the Middle Ages* (1995). The Church was a major force in promoting technology, especially as it was used by the Cistercians; *The Cambridge Companion to the Cistercian Order* (2012) includes material on architecture, agriculture, economics and libraries. Agriculture was the foundation of medieval life and a very detailed examination of agriculture can be found in Grenville G. Astill and John Langdon's *Medieval Farming and Technology: The Impact of Agricultural Change in Northwest Europe* (1997). Lynn White's *Medieval Technology and Social Change* (1962) remains one of the foundational texts on the period. One of the key technologies that transformed society in this period was printing. A broad overview can be found in S. H. Steinberg and John Trevitt, *500 Years of Printing* (1996), while Elizabeth L. Eisenstein's *The Printing Press as an Agent of Change: Communications and Cultural Transformations in Early-Modern Europe* (2009) remains the leading text on the subject.

7 The Industrial Revolution and the Rise of European Power

Financed with gold and silver from the Americas and supported by slave labor, Europe shifted to a cash economy. One of the major economic developments was the creation of the Bank of England, the first national bank. It promoted commerce by stabilizing the currency and protecting investments. This allowed for the growth of industry as more people invested in the growing capital system. With more money and risk spread out by the creation of companies, a great wave of industrialization was started. The first major sector to be affected was the textile industry as inventors came up with new devices to mechanize spinning and weaving. The changes in the textile industry had the biggest impact on Britain because textiles were not as strongly controlled by guilds as they were in other parts of Europe, and the parliamentary system allowed mill owners and investors to influence government regulation. The invention of a practical steam engine and the introduction of the division of labor combined to create the first factories. The rapid changes in technology led to conflicts and social problems, but Britain became the largest empire in history on the strength of the Industrial Revolution.

> And did the Countenance Divine,
> Shine forth upon our clouded hills?
> And was Jerusalem builded here,
> Among these dark Satanic Mills?
>
> William Blake, *Milton: A Poem*, 1804

The second stanza of William Blake's poem introduced the term "satanic mills" and it was widely believed that it was a reference to the factories that were springing up across Britain. The vision of the diabolical, dark and dangerous mills was further reinforced when Robert Bridges, England's Poet Laureate, asked Charles Hubert Hastings Parry to set the poem to music as a hymn that would be used to promote patriotism during the First World War.[1] When Blake was writing his epic poem, England was in the midst of one of its greatest periods of challenge, with industrialization at home changing the economy and social relationships, the recent loss of the American colonies, and Napoleon being crowned emperor of France. In 1803, Thomas Malthus published the second edition of his *Essay on the Principles of Population* that argued that birth rates, if not restrained in some fashion, would increase exponentially, while at best food production could rise arithmetically, leading inevitably to competition for resources, famine and social anarchy. This was also a commentary on Catholics, the Irish and the poor, whose birthrates were high, causing some in the English middle and upper classes to fear moral decline and rebellion. The factories were making a new class of people wealthy, but also creating urban slums. Some workers rebelled against the new economy and took up arms against the machines.

Herein lies the great historical controversy of the Industrial Revolution. For those at the bottom of the social hierarchy, the move away from an agrarian life to an urban one was often devastating, and yet it made Britain powerful and wealthy enough to create the modern industrial world and its attendant material benefits. For those able to take advantage of industrialization, they were made so wealthy that they lived like ancient emperors, surrounded by luxury and waited on by armies of servants. In contrast, the slums of the industrial era were truly squalid: dark and dangerous, with open sewers, home to disease, alcoholism, drugs and violence. Work was at the discretion of the factory owners and other employers, so working conditions were often poor. Work was often dangerous, but it was less expensive to replace injured workers than to buy safer equipment. Terms like proletariat, wage-slaves and underclass all refer to the workers in the mills and factories, while the owners were robber-barons, profiteers and fat cats.

Great as the division between rich and poor was, and rigid as the class system could be, there were astonishing social movements made possible by industrialization as well. Efforts to reject slavery, institute civil rights and end child labor, and the suffrage movement's efforts to get the vote for women, were part of the era. Most of the new industrial states either were or became democratic states. Bad though the bad parts could be, industrialization led to the most significant revolution in class structure in history.

The distinction between rich and poor was evident in the new industrial states, and this led to a number of thinkers writing about the effect of industry on human relations, politics, law and history. The best known of the industrial-era thinkers was Karl Marx, but the list of new philosophers of the industrial age includes Adam Smith (economics and government), Thomas Malthus (population), David Ricardo (economics), Jeremy Bentham (utilitarianism), John Stuart Mill (economics and the social contract), Friedrich Engels (economics and socialism) and Thomas Jefferson (politics), to name just a few. All of these thinkers tried to make sense of the world in the rapidly changing circumstance introduced by revolution and industrialization.

Despite the fear of the factory as an agent of change to the social traditions of the country, the terrible working conditions of the mills and mines, and the squalor of the slums that grew up around the factories, Britain emerged from the period the most powerful empire the world had ever seen. Its territories stretched farther than the Roman or the Mongol empires at their heights, while Britain's navy commanded the oceans and her merchants traded globally. The Duke of Wellington is reputed to have said that the Battle of Waterloo was "won on the playing fields of Eton," but it would be far more accurate to say that the Battle of Waterloo was won on the factory floor, as every British boot, ship, bullet, tent or cannon could be manufactured faster, have superior quality, and cost less than they could be produced in France or anywhere else.[2]

The Industrial Revolution is often portrayed as the triumph of mass production and the replacement of muscle power by mechanical power. Yet from a historical point of view, both of these had existed for generations, as the example of the mass production at the brickworks of ancient Egypt and use of the multiple waterwheels at the Roman mill at Barbegal indicate. Rather than simply the concepts of mass production and mechanical power, there were two crucial modifications that came together to produce the revolution in manufacturing. The first was the introduction of the division of labor to mass production, and the second was the invention of the steam engine as the primary source of energy for mechanical power. The division of labor had increased in the proto-industrial period as the number of skilled and semi-skilled workers increased and they tended to specialize in particular activities such as spinning yarn or making nails. Division of labor allowed for the "de-skilling" of manufacturing so that a worker needed to learn only one skill or a small set of skills rather than master all the elements in a production system. Steam engines dramatically increased the power available for manufacturing and allowed production to be centralized and located near markets, rather than on watercourses that tended to be far from urban centers. These two developments came together to increase the speed of production, as workers became more efficient and machines automated tasks that had previously been done by hand.

Banking and Patents

In addition to the new inventions, there were two important invisible technologies that helped make the Industrial Revolution possible. Both had their roots far earlier than the eighteenth century, but they became part of the institutional framework for industrial development. They were the creation of the first modern national bank, with the founding of the Bank of England, and the establishment of patent law.

Various forms of banking had existed back into ancient times, and the introduction of coinage and later paper money facilitated trade and investment. Banking was usually a private business and tended to be used only by the very wealthy and the leaders of states. The largest European bank prior to the eighteenth century was the Medici Bank that operated from 1397 until it failed in 1494. At its height, the Medici Bank had branches across Europe and had financial interests through trade running from England to the Silk Road. A large part of the failure of the bank was loans it made to leaders to pay for wars or luxurious lifestyles. These loans were often forced on the bank by the threat of restrictions on the trade interests of the Medicis, and there was no way to collect the loans if a king refused to pay. The failure of the Medici bank set off an economic collapse in Europe that set back trade and innovation for many years.

In the seventeenth century, King William III of England was desperate for money, so William Paterson (1658–1719) proposed to loan the king £1.2 million by creating a company of investors. In 1694, a royal charter was granted to the Governor and Company of the Bank of England. In large part because the government of Britain was held jointly by the monarch and parliament, money issues had to be handled in public. The Bank of England was not a general bank, but was designed solely to lend money to the government and manage government debt. Some of the investors in the Bank were members of parliament or in a position to influence members. This meant that the government had a vested interest in seeing loans repaid. Repaying loans became more important as the concept of national debt (rather than private debt held by the Crown) led to efforts to maintain the value of the currency by holding reserves of bullion. Thus, coins and paper money became tokens of exchange that had value based on the reserve rather than the material (gold or silver) of the coin itself. Because of the Bank of England, the currency of England was the most reliable in Europe. This was further enforced by the Bank Charter Act of 1844 that gave the Bank the sole right to issue banknotes, and guarantee the value of the British pound by holding gold reserves.

Although having a national bank did not eliminate all financial problems, it tended to stabilize one of the largest areas of financial transactions – that of the public with the government. Over time, the Bank of England encouraged investment in three ways. First, by stabilizing the currency, people were encouraged to loan money and thus spread the risk and benefits of investment. Second, by issuing bank notes (paper money) commerce was facilitated. Third, the Bank of England also operated as a commercial bank and became in effect the banker for other private banks that were set up to finance business and eventually opened banking to the public. In combination with the charted companies, the Bank of England gave the British a financial advantage over her European competitors.

Inventors and investors might be able to get financing because of the bank system, but making a profit meant more than simply being first to market. It also meant controlling access to the technology to maximize income. Monarchs have often given individual "letters patent" granting some form of monopoly of production to favored people, but the practice became more codified as European states began to see the benefit of technological innovation. One of the earliest efforts to offer monopoly protection was in 1474, when the Republic of Venice issued a decree that devices brought to the market and their details communicated to the government could gain legal protection against anyone copying the device. In 1623, England enacted the Statute of Monopolies, granting protection for fourteen years.

VI. Provided alsoe That any Declaracion before mencioned shall not extend to any tres Patents and Graunt of Privilege for the tearme of fowerteene yeares or under, hereafter to be made of the sole working or makinge of any manner of new Manufactures within this Realme, to the true and first Inventor and Inventors of such Manufactures, which others at the tyme of makinge such tres Patents and Graunts shall not use, soe as alsoe they be not contrary to the Lawe nor mischievous to the State, by raisinge prices of Commodities at home, or hurt of Trade, or generallie inconvenient.[3]

By the beginning of the eighteenth century, it was a requirement that those seeking a patent had to write a detailed description of the device, including the manner of operation. These "specifications" became the foundation for litigation regarding the granting of patents, so that Richard Arkwright lost his patent protection for his yarn spinning machine because it seemed to have been based on designs created by others, and thus the specifications in his patent description were too general and included material already in the public domain.

Although specifications helped make patents clearer, it was not always clear who in the government was really responsible for patents. Inventors sought the help of various departments and their member of parliament, while others went to the court of the King, or the Admiralty. In all, seven offices had to look at any patent, and each stage required a fee. Patent problems, such as Arkwright's loss of patent protection, were decided by parliament. This led to many delays and prevented people from seeking patents if they lacked the money or social contacts. In 1852, the Patent Office was created. In addition to making the patent process simpler, it also reduced the fees and a patent was good for all of the British Empire, not just Great Britain.

The power of the patent as an instrument of political policy is perhaps best seen in the development of the American patent system. The US Constitution contained a specific clause covering copyright and patents. It gave the government the power to grant monopolies to creators and inventors: "To promote the Progress of Science and useful Arts, by securing for limited Times to Authors and Inventors the exclusive Right to their respective Writings and Discoveries."[4]

The original Patent Board was replaced by a separate Patent Office within the State Department in 1802, and by 1836 there were more than 10,000 patents issued by the American government. Some of these were copies of devices created in other countries and patented by unscrupulous people evading patent controls. This situation required inventors of major patents to file in many countries in an effort to protect their work in major markets. It often ended up with a patchwork of coverage that led to a great deal of litigation. The Paris Convention of 1883 brought in the first international patent law, and eventually patents gained in one country were accepted by all the signatories.

⚙ Textiles

Some of the most significant patents and patent fights were for inventions in the textile industry. Textile production was both the engine of change and the financial foundation for that change. The popular images of the Industrial Revolution tend to be dramatic pictures of steel mills, locomotives and coal mines, as seen in J. M. W. Turner's painting "Rain, Steam, and Speed – The Great Western Railway" (1844). In reality, it was the weaver's loom that was primarily responsible for setting off the revolution.

Textiles are one of the oldest manufactured products and have been one of the most profitable for the simple reason that everyone needs clothes. And because clothes wear out, the demand never goes down. In some parts of Europe, the weaving guilds were so wealthy that they financed major public works projects such as churches, roads and clock towers. By the beginning of the eighteenth century, the manufacturing of cloth in Europe was highly specialized and one of the most skilled jobs around. There were four major types of textile material: wool, linen, cotton and silk, in order of their value from lowest to highest. To understand why textile technology developed the way it did, it is useful to understand the general process that each textile underwent to go from the field to the finished product.

Wool textiles, coming from sheep, one of the earliest domesticated animals from around 9000 BCE, have been a major industry from prehistoric times. Sheep were sheared, generally in the spring (they needed a thick coat for winter and then it would regrow during summer and fall), and the fleeces were cleaned. The individual hairs then had to be arranged so they went in the same direction. This activity, called "carding," was done by a kind of combing or brushing. This produced a loose mat of fibers, that then had to be turned into thread by twisting the fibers together. This was done by hand. Because of the natural curl of the wool, these threads were very strong. The threads could, at this point, be dyed, using natural dyes such as woad (also known as glastum) or indigo. Threads could then be woven as fine cloth, or several threads combined to produce heavier yarn used for hand knotting or later knitting.

Linen production had several elements in common with wool production, but the material comes from the flax plant, rather than an animal. The flax was harvested and the seeds separated for other uses. To separate the fibers from each other and the woody material in the stem, the stems were left in the field, or soaked in water. This process, called "retting," was followed by some form of crushing that freed the fibers. The fibers, as for wool, had to be combed, often by beating a large handful of stems against sharp pins and drawing them through the pins until they were separated and aligned. The fibers were then twisted together into thread or yarn, dyed if desired, and then used for textile production.

(a)   (b)

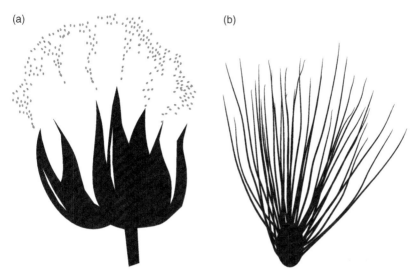

Figure 7.1 Cotton and cotton seeds. The cotton pod contains many seeds and each seed has multiple threads of cotton fiber.

The fiber source that would help propel new invention in the textile industry was cotton. Cotton is a global plant, with wild varieties found from Peru to Australia, but commercially useful species, especially *Gossypium hirsutum*, grow best in warm climates. The earliest cultivation of cotton was in India in the Indus region some 7,000 years ago. In the Americas, it was being grown in Mexico by 3000 BCE. In Europe, cotton was a luxury good, imported first from Egypt and India. It was originally an expensive fabric because it was very labor intensive to make, and could only be grown in places such as Egypt and India, far from the British textile industry. Cotton fibers are firmly attached to the seed and bound up in the cotton pod (Figure 7.1). Each seed had to be pulled free from the fibers by hand. Once the fibers were freed, they were combed and twisted into thread or yarn.

The most amazing textile, and the most expensive, is silk. Silk comes from silkworm cocoons (primarily *Bombyx mori*). These highly specialized larvae were first used as a textile source around 2700 BCE in China. The silkworms live on mulberry leaves, and then spin a cocoon as part of their life cycle. The cocoons are soaked in hot water (one of the origin stories has a cocoon falling into a cup of tea), and each cocoon, made of a single strand of silk, is then unwound. The fibers are aligned and spun into thread for weaving.

Once the threads or yarn were produced, they could be woven or knotted together to make cloth. The basic weaving system was the same whether it was developed in Asia, Africa or South America. One set of threads was arranged vertically, and then a second set was laced horizontally over and under the vertical threads. To create patterns, yarn

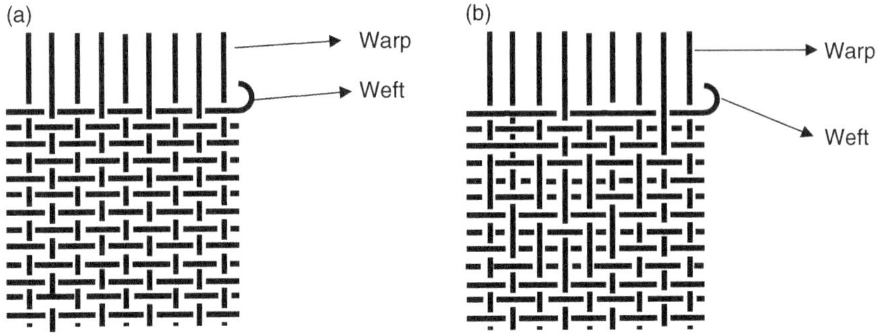

Figure 7.2 Basic weave and pattern weave. In a basic weave, the threads follow a simple over and under distribution. In a pattern, threads are passed over or under multiple threads to reveal more or less of the color of the thread, forming a more complex design.

of different colours was added to the weave. Over time, these patterns became more and more complex (Figure 7.2).

At the beginning of the Industrial Revolution, each of the separate steps in fiber production was done by hand or small hand-operated devices such as the spinning wheel, first invented in China and making its appearance in Europe around 1280 CE. Textile production was partly a home-based activity, but over time almost every civilization had guilds of dedicated craftspeople who produced large quantities of cloth. Weavers, who needed bulky and increasingly complicated equipment and years of training, were highly skilled workers. Prior to the Industrial Revolution, weavers in the Middle Ages and the early modern period were often powerful and wealthy, and the weaving guilds financed development including roads, canals, churches, temples and schools.

The introduction of new machines changed the nature of textile manufacturing and had a major impact on British society. The system most widely used across Europe was the "putting-out" system or the workshop system. The merchant (sometimes known as a "draper") arranged for the raw materials such as wool or linen to be delivered (put-out) to the workers, who carded it, spun the yarn and either delivered the yarn to the merchant to pass on to a weaver, or wove the fabric themselves on a handloom. This was often a family activity, with children carding and older people spinning and weaving. Both men and women spun and wove cloth, although in Europe weaving as a profession increasingly became dominated by men as more equipment was introduced. In other parts of the world, weaving was often a female occupation.

The merchant then collected the textiles, paying by the piece with a sliding scale for quality. In many cases there was further processing such as dying and fulling. Fulling was used for woolen cloth and it cleaned and thickened the cloth. The process began with scouring, when urine (traditionally human) was used to clean and bleach the cloth. The ancient method, which was used by the Romans, was to put the cloth in tubs or basins and beat it by hand, walk on it or pound the cloth with clubs. In the Middle

Ages, European fullers began adding "fuller's earth" (a type of clay containing hydrous aluminum silicate) that de-colored oils and made cloth whiter. The pounding spread the fibers out (known as "felting"), increasing the volume of the cloth and making it softer and more water resistant. The liquid was cleaned out with water and the fabric stretched on large frames. As textile production increased in Europe, hand fulling was replaced by the use of fulling mills that employed cam-driven hammers to pound the cloth. Fulling was such an important job that the Anglophone names Tucker, Walker and Fuller are derived from the job.

This production system was very popular with weavers, who could work as much or as little as they wanted, could negotiate their rates, and could set the standards for the quality. They were masters of a small business and that carried social status. Local weavers often formed collectives, some simply loose associations that helped each other, while others, such as the silk weavers, formed powerful guilds.

Market pressure and rising demand fostered innovation in the textile industries. One of the first changes was the gradual introduction of larger looms, until the width of the loom made it difficult to pass the weft (horizontal) thread from side to side by hand. The solution was to use a shuttle to pass the yarn from side to side rather than weaving it by hand. This increased the speed at which textiles could be produced. It is not clear when shuttles were first used. It is likely an Asian invention from around 500 BCE based on using a stick to push the weft thread, but various forms of shuttles were used in Greece, Rome, Sri Lanka and China in ancient times. By the seventeenth century it was widely used in Europe and Asia. The most efficient form was the boat or bobbin shuttle, constructed so that thread wound around a dowel or bobbin was placed in the middle (Figure 7.3). The thread was then pulled through a hole in the side of the shuttle. The bobbin could be removed to replace the thread or change colours.

Looms became increasingly complex. Two of the most important innovations were the introduction of the moving heddle and the batten with reed comb. The heddle is a ring that the warp thread passes through, allowing the thread to be moved up or down by a foot-operated treadle rather that the weaver moving the weft thread over and under each warp thread by hand. In simple weaving, half the warp threads would be up and the other half down. The shuttle was then passed from one side to the other.

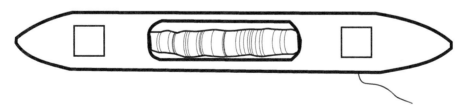

Figure 7.3 Top view of a boat or bobbin shuttle. The weft thread is on a spool in the middle and feeds out through the side of the shuttle.

The position of warp threads would then be reversed and the shuttle sent back, forming the next row of fabric. With this device added to the loom, more complex patterns were possible, since warp threads could be controlled individually or in groups.

In hand weaving, once the weft thread was passed from one side to the other, it had to be beaten down with a comb so that it was as close to the previous line of thread as possible. The batten and reed comb was a set of strings or teeth (thin strips of wood or metal that looked like reeds) attached to a pivoting arm that spanned the width of the loom and pressed the weft threads down all at once. This was faster and more consistent than using a hand-held comb.

Using the shuttle and heddle on larger weaving frames was far more efficient than hand weaving, and weavers ran into the problem that they could weave faster than yarn could be produced, and thus a kind of inter-tool race developed between yarn production and weaving.

As demand for textiles rose through the eighteenth century, the cottage industries run by the weavers did well as demand rose faster than weavers could keep up. This created a tension between those who saw innovation both as a way to overcome problems in the system of production and as a way to capitalize on the growing demand for textiles, and the weavers whose income increased using the established system. Most of the efforts to improve textile production were rejected by the weavers because they threatened profits and challenged the skilled status of the workers. One of the first examples of this struggle was the flying shuttle, introduced by John Kay of Bury, England. It was a simple modification of the existing shuttle. Rather than slide the shuttle from side to side by hand, which limited the width of fabric to the arm length of the weaver, the flying shuttle was propelled by a sling, and later various systems of springs. This increased the width of cloth and the speed of weaving. It was seen by weavers in both England and France as an attack on their livelihood, because by increasing the amount of cloth that could be produced in a certain length of time, it decreased the value of each piece. In England, Kay was physically attacked by weavers. Although skilled weavers did not like the flying shuttle, other textile manufactures did, but John Kay and his partners did not profit by his invention because it was pirated by other manufacturers who refused to pay royalties. At one point there was even a consortium of patent pirates known as the "Shuttle Club" who agreed to pay each other's legal bills fighting Kay's patent infringement cases. Kay eventually moved to France and contributed to the mechanization of French weaving, but died in relative poverty.

Despite the introduction in the sixteenth century of the treadle-powered spinning wheel that speeded up the production of yarn many times over hand spinning, by the beginning of the eighteenth century weavers could weave faster than carders and spinners could produce yard. Yarn production became the bottleneck in textile production, so it was the logical target of innovation. The roller spinning machine was invented by Lewis Paul (c. 1700–59) and John Wyatt (1700–66), the hand-driven carding machine

Table 7.1 Key inventions for the textile industry.

Date	Invention	Inventor*
1733	Flying shuttle	John Kay (of Bury)
1738	Flyer-and-bobbin	Lewis Paul and John Wyatt
1738	Roller spinning machine	Lewis Paul and John Wyatt
1748	Hand-driven carding machine	Lewis Paul; other versions by Richard Arkwright and Samuel Crompton
1764	Spinning jenny	Thomas Highs; also James Hargreaves
1769	Spinning frame	Thomas Highs, acquired by Richard Arkwright with John Kay (of Warrington)
1771	Water frame	Richard Arkwright
1775	Carding engine	Richard Arkwright
1779	Spinning mule	Samuel Crompton
1784	Power loom	Edmund Cartwright
1790	First steam-powered mill	Richard Arkwright
1803	Dressing frame for continuous weaving	William Radcliffe
1804	Jacquard automated loom	Joseph Marie Jacquard

* Debate about the original inventor versus the first person to patent a device clouds the issue of who should be credited with the first working model.

by Richard Arkwright (1733–92), the spinning jenny by Thomas Highs (1718–1803) and perfected by James Hargreaves, and finally the spinning frame invented by Thomas Highs and acquired by Richard Arkwright. Highs is a bit forgotten as an inventor, in part because he either did not or could not defend his invention, so others utilized and improved on his designs (see Table 7.1).

✿ Richard Arkwright: Invention and Social Mobility

Richard Arkwright serves as a case study of the inventive society that began to appear in eighteenth-century Britain. Arkwright's father was a tailor, a skilled trade, but the family was not wealthy enough to send him to school, so he was taught to read by a cousin. He began his professional career as a barber and wigmaker, but became interested in carding and spinning cotton, which was at that time entirely done by hand. With the help of the clockmaker John Kay (fl. 1733–64, not the inventor of the flying shuttle), Arkwright succeeded in making a mechanical spinning machine to turn raw cotton into thread. In 1769, he patented the spinning frame, later known as a water frame when it was powered by waterwheels. Arkwright was extremely aggressive in his work to design and patent textile equipment, and had as a plan to integrate all yarn production from raw material to finished yarn in a single location. By mechanizing the process of yarn production, he did two important things: he deskilled yarn production, and he made it possible to apply power to the spinning process, first using animal power, but quickly using water power.

Arkwright attracted investors, including Jedediah Strutt (1726–97), who was already in the textile business making stockings. The stocking frame was a mechanical device that imitated the hand knitting process and could create tubes of fabric. Strutt's contribution was the Derby Rib machine that put ribbing in the fabric (effectively a purl knit), making it more interesting to look at and more elastic for a better fit. Silk was the most desirable stocking material, but it was expensive, so cotton was a good alternative, particularly once Arkwright's water frame made cotton thread much less expensive.

Arkwright opened his first big mill at Cromford in Derbyshire in 1771. This was not the first spinning mill, but it was the first really successful cotton spinning factory. His machines and system were copied, with similar mills appearing in Germany and the United States. The Cromford mill has been preserved as a historical site.

Spinning cotton gave Arkwright power and wealth, but all was not perfect. In 1779, one of Arkwright's mills at Birkacre, Lancashire was destroyed in an anti-factory riot. He attempted to obtain a patent in 1775 that covered all aspects of spinning mills and would have given him an almost complete monopoly on the spinning industry. Despite his efforts, Arkwright's patents were eventually nullified by parliament in 1785, partly because he had borrowed ideas from other inventors such as Thomas Highs. Regardless of these setbacks, his inventions and the size of his industrial empire ensured him a place in history. Arkwright was knighted in 1786.

The story of Arkwright reveals some emerging trends in technology in the late eighteenth century. There was enough social flexibility in Britain that Arkwright could benefit from his invention, but unlike John Kay's shuttle, Arkwright had a vision of his invention within a bigger system. He was also more successful in gaining financial support and making commercial and social connections. Both Arkwright and Kay had problems defending their patents, but with better plans the invention became less important than the system that grew up around it. Arkwright's story would also contribute to a growing cultural ideal of the "self-made man" – someone who comes from low standing and by cleverness and hard work rises to a position of high status and wealth.

Ralph Mather (c. 1755–1836), a preacher who objected to the social destruction brought about by textile industrialization, published *An Impartial Representation of the Case of the Poor Cotton Spinners in Lancashire*, saying:

Arkwright's machines require so few hands, and those only children, with the assistance of an overlooker. A child can produce as much as would, and did upon an average, employ ten grown up persons. Jennies for spinning with one hundred or two hundred spindles, or more, going all at once, and requiring but one person to manage them. Within the space of ten years, from being a poor man worth £5, Richard Arkwright has purchased an estate of £20,000; while thousands of women, when they can get work, must make a long day to card, spin, and reel 5,040 yards of cotton, and for this they have four-pence or five-pence and no more. (Mather 1780)

✿ The Growing Textile Industry

The perfection of the spinning mills produced vast quantities of yarn at very low cost. By the 1790s, there were thousands of spinning machines in Europe, and Arkwright himself employed more than 20,000 people. Now it was the weaving end of the industry that received the attention of the inventors. A somewhat unlikely source for mechanical advancement was Edmund Cartwright (1743–1283), an Anglican clergyman. In 1784 he invented his first power loom, and he patented it in 1785. This loom mechanized the process of raising and lowering the warp threads and sent the shuttle back and forth automatically. Cartwright set up a factory in Doncaster, and continued to work on the very complex problem of weaving. Some of these problems he solved, introducing methods to weave simple patterns, but he was not so successful as a businessman and his mill was taken over by his creditors in 1793.

The perfection of the basic weaving machine was achieved in 1789 when William Radcliffe (c. 1761–1842) opened a cotton weaving factory in Mellor, Cheshire. He added an important modification in 1804 when he created a ratchet wheel that moved the woven cloth forward automatically while the weaving machine was working. In theory, weaving could be made infinite with these new innovations, as the yarn for weft and warp could be added while the machine was running and the woven cloth taken off without end.

Basic fabric production supplied a large portion of the textile market, but weaving was part of fashion, not just utilitarian need. As such, weaving patterns were one of the most important aspects of the industry and made the process much more complex than simply setting warp threads and then crossing them with the weft threads. To produce patterns, especially the intricate ones found in silk brocades, required many different colours of yarn to be woven into the cloth. This meant that each warp thread had to be controlled individually. In the oldest versions of pattern weaving, the weavers simply pulled up or left the warp threads as they pulled the shuttle containing the weft thread across the loom. As the pieces of cloth got larger and the patterns more complicated, the weaver had to have one or more assistants, often children, who would pull up the warp threads by pulling on a cord with a ring on the end (heddle) through which the warp thread passed. This was done according to a set of instructions to complete the pattern. It was slow and subject to many errors.

The Automation of Weaving

In 1725, the French silk weaver Basile Bouchon invented a system to control the warp-lifting cords using a piece of paper tape with holes punched in it. This was probably the application of a system that had been developed to automate the ringing of clarion bells that used pegs set in a cylinder. As the cylinder rotated, the pegs triggered the bells in a set order. The bell master would use a sheet of paper with holes in it to set the pegs for different melodies. In the loom, the holes punched in the paper controlled whether the warp thread was raised or not.

Bouchon's system was useful, but it was almost as labor intensive as the old system. In 1728, one of his co-workers, Jean Falcon, substituted punched cards for the paper tape. The cards were more durable and could be joined together to make a larger pattern. This innovation was further developed by Jacques de Vaucanson (1709–82). Vaucanson was a master mechanic, well known for his automatons such as a duck that ate grain, drank water, flapped its wings and defecated. This experience gave him great insight into how to control machines. In 1741, Vaucanson was appointed as the inspector of silk manufacturing, and had the job of both inspecting and reforming the industry in order to compete better with English and Scottish weavers. He took the ideas of Bouchon and Falcon and created a model of an automated weaving machine in 1745, but his ideas were not well received by the weavers (and at one point he was pelted with stones), so they were not introduced to the industry.

Joseph Marie Jacquard (1752–1834) was a silk weaver who saw the advantages of automation. He took the ideas of the earlier inventors, especially Vaucanson's, and created the first automatic pattern weaving loom. In addition to the punch cards to control the warp threads, he added a ratchet mechanism so that as each weft thread was laid down and drawn into place, the punch cards would advance one row. In 1801, Jacquard showed his automated loom at the Industrial Exposition in Paris. Three such "Napoleonic" Expositions were designed to promote French industry and press for modernization so that France could compete with the increasingly industrial Britain. Many French weavers feared the Jacquard loom just as the English weavers feared the water frame and the power loom, but the advantages of Jacquard's loom were so great that it was declared public property in 1806, and by 1812 there were more than 10,000 Jacquard looms in France. Jacquard was rewarded for his work with a government pension and a royalty for each machine.

Despite the brilliance of the Jacquard loom, it was not in France that it reached its greatest use. The elaborate patterns were seen as both royalist and too expensive for Revolution-era France. The technology was taken up in Britain and contributed to the power of the British textile industry, as patterns such as argyle and paisley became part of fashions that could be afforded by all the people.

⚙ The Luddites

The creation of the spinning mill and the weaving mill undermined cottage industries and began to put weavers out of business. The beginning of the nineteenth century was a period of economic strife in Europe, as the French Revolution and the rise of Napoleon led to a long period of war. Weavers had acted to protect their work for generations, going back as far as the Middle Ages, so when mills were opened in Nottingham, the local weavers banded together and began to attack and break the hated "frames." A wave of such actions occurred in 1811, and the weavers organized themselves, claiming to follow a "King Ludd" (or sometimes "General Ludd").[5] The Luddites, as they came to be called, destroyed wool and cotton mills in Nottinghamshire, Yorkshire and Lancashire, the heartland of the mill counties where plentiful water courses made water power easily available (Figure 7.4). There was popular support for the Luddites in the region, but in part because of the concern about civil unrest during a time of war, the government passed the Frame Breaking Act in 1812, making frame breaking a capital offense; 12,000 soldiers were sent to enforce the law. At least twenty-three people were sentenced to death and others were transported to Australia for their participation.

As the textile industry expanded and became more profitable, there were changes not only in the status of the textile workers such as the weavers and their system of cottage production, but in the entire chain of production. A long series of Enclosure Acts were passed by the British government between 1750 and 1860. The enclosures transferred control of common land, where many people had the right to graze animals, to private hands. Small farmers, usually renters, were then turned out or could not survive without access to the common land. By 1845, some 28,000 km^2 (7 million acres) had been enclosed. A large portion of this land was grazing land and had been used to raise sheep for the textile industry. Many of the displaced people moved into the cities and mill towns seeking work.

Once the workers were in the factories and cities, they had few amenities, lived in great poverty, and were often treated as being less valuable than the equipment they tended. Child labor, the exploitation of women, the use of corporal punishment on the shop floor, and no social support added to the danger of the workplace. One of the ways that owners exploited the workers was by the truck system. In this system, the owners paid the workers in products, often at inflated rates, or paid in tokens or promissory notes called "tommy tickets" that could only be spent at a company-run store, where again the prices were inflated. This caused unrest in many regions of England and led

WHEREAS,

Several EVIL-MINDED PERSONS have assembled together in a
riotous Manner, and DESTROYED a NUMBER of

FRAMES,

In different Parts of the Country:

THIS IS

TO GIVE NOTICE,

That any Person who will give Information of any Person or Persons
thus wickedly

BREAKING THE FRAMES

Shall, upon CONVICTION, receive

50 GUINEAS

REWARD.

And any Person who was actively engaged in RIOTING, who will
impeach his Accomplices, shall, upon CONVICTION, receive the
same Reward, and every Effort made to procure his Pardon.

☞ Information to be given to Messrs. COLDHAM and ENFIELD.

Nottingham, March 26, 1811

Figure 7.4 Anti-frame-breaking poster offering a reward for turning in people associated with the
Luddite movement.

to the introduction of the Truck Act 1831, that ruled that workers had to be paid their
wages in "coin of the realm." Unlike some efforts to ameliorate working conditions by
acts of parliament, this one had support from a number of mill and factory owners,
who reasoned that the truck system created a kind of factory monopoly that prevented
workers from buying their products in an open market.

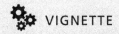

Neo-Luddites: Fighting the Future or Finding Another Path?

We have always had a love–hate relationship with technology. Loving technology is easy – just buy the latest version of almost anything (especially if you don't actually need the new thing) and you are supporting and participating in the technological world. Expressing concerns about technology is more difficult. At what point do everyday concerns about things like having our Facebook privacy circumvented or worrying that we might lose our job to a robot cause people to form an anti-technology worldview? It may be socially acceptable to say that you won't buy a driverless car or don't use Twitter, but objecting to technological society as a whole is often seen as old fashioned, naïve or dangerous. Yet there are a growing number of people who are saying that they have had enough of the technological world.

The term "neo-Luddite" is a general term for a broad collection of individuals and groups who oppose the use and/or further development of modern technology, often using the Industrial Revolution as the transition from controllable to uncontrolled modern technology. Some people object to the term because it refers to the failed violent Luddite movement of 1811, while others embrace the term because it connects them to a long history of resistance to technology displacing people and undermining human interests. If there is anything that unites the varied views on technology it is that the future will be bad if the world continues to rely more and more on technology, and conversely, the rejection of technology will make us happier, healthier and more socially connected.

Many neo-Luddites trace their ideas to Jacques Ellul, who published his critique of modern technology in *The Technological Society* (1964). Ellul argued that the logic of technology was based on efficiency of production and the subordination of the natural world while ignoring all human concerns. A number of the most prominent figures in the anti-technology movement such as Kirkpatric Sale, Chellis Glendinning and John Zerzan agree with Ellul's critique. Following this idea, most neo-Luddites propose a return to a mostly pre-Industrial Revolution and agrarian level of technology. Sale is an advocate of autonomous regions while Glendinning is a proponent of bioregional land-based culture. Zerzan's position goes further and has been described as "anarcho-primitivism" because he believes that we should return to a pre-agricultural culture and resume the hunter-gatherer lifestyle.

One of the biggest differences among neo-Luddites is about what individuals, groups and governments should do to counteract the effect of technology. Most anti-technologists promote the idea of individual responsibility combined with some level of intervention, either through markets or by governments to protect the environment and reduce social inequality. Some anti-technologists believe that only large-scale intervention has any chance of preventing future disaster. A smaller group believe that nothing can be done and that only a global collapse can lead to a technologically responsible Civilization 2.0. Sale, for example, had a bet with Kevin Kelly (former publisher of the *Whole Earth Review* and founding editor of *Wired* magazine) that by 2020 there will be a global catastrophe involving environmental disaster, warfare between the rich and poor, and the collapse of global currency.

Most anti-technology proponents reject violence because it is morally wrong and likely to alienate the general public, but the most extreme supporters of anti-technology ideas argue that only direct action can force people to change their behaviour. The most notorious example of direct action was by Theodore Kaczynski, better known as the Unabomber. Kaczynski was a mathematician who left a job at Harvard University to live in the wilderness in Montana. Starting in 1978, he sent sixteen bombs to various people, killing

three and wounding twenty-three. In 1995, he wrote to a number of media outlets saying that if his anti-technology essay "Industrial Society and Its Future" was published in a major newspaper, he would not send any more bombs. It was published in the *New York Times* and the *Washington Post*. The FBI arrested Kaczynski in 1996.

Anti-technology ideas are important to consider when discussing the history of technology. They are part of the historical record, but they also remind us that technology is a product of human activity and no human activity is above critical examination.

❧ Reformers

Not all mill owners were unsympathetic to the conditions of the workers, recognizing both the humanitarian problems of poverty and the economic cost of a workforce made feeble and unreliable by illness, alcoholism and violence. The leading example of an effort to turn the factory system into something more humane was Robert Owen (1771–1858) and his mill at New Lanark, Scotland. The New Lanark mill, utilizing the water power of the falls on the River Clyde, was started by Dale and Richard Arkwright. Owen persuaded his partners to buy the mill and he began to run it following philanthropic principles, such as not allowing child labor, and offering infant care and education to the children of the workers. Although the mill continued to be profitable, it was not making as much money because of the cost of Owen's improvements. Owen formed a new investor group in 1813 that included the political thinker Jeremy Bentham, and leading Quaker William Allen.

Although New Lanark was a commercial success, two other ventures in a kind of social/egalitarian commerce, at Orbiston near Glasgow and at New Harmony, Indiana, both collapsed. Owen ended his connection with New Lanark in 1828, and turned his attention to promoting better working conditions, socialist politics and the co-operative movement. While Owen's ideas about a more egalitarian workplace did not change the life of many people in his lifetime, many of the things he promoted, such as free education for the workers, inspectors to ensure that factories were safer and that owners followed the laws regarding employment standards and wages, and the reform of the Poor Laws to offer some relief to the poorest of the poor, remain part of his legacy.

❦ The Factory System

The textile industry started as a way to overcome the limitations of hand work. In the process, it made clear the power of the factory system. First, it centralized work, getting rid of the less economical putting-out system. It was far more efficient to bring raw materials to a single site than to send it out to many small processors.

Second, it improved the overall quality of goods. Although the very best fabrics were the work of highly skilled artisans (and this remains true to the present), the general quality of textiles was low until mass production became common.

Third, it deskilled work. The worker in the textile factory had to learn only a few activities and this was faster and less expensive than training an artisan. While this was one of the major objections that skilled workers had about the factory system, it also made mass employment possible. The deskilling of the production system was matched by the rise of a new class of skilled workers, the technicians and engineers who designed, built and maintained the automated equipment. It also created managerial and office work, as people had to be employed to oversee production, manage supply and delivery of materials, keep track of work time and payrolls, do accounting, pay taxes and do all the other things associated with commercial ventures.

Fourth, the creation of the factory was made possible by a changing economic system that favored investment in a generally open market. There was a promotion of innovation as different groups could raise the capital to improve their machines and hence their competitive position, or by collective investment begin new ventures.

As the factory system was perfected, there was another problem that the mill owners needed to address. That was how to create and maintain a market for their goods. It was important to produce goods, but those goods had to be sold to someone, and that proved to be a different kind of technological and social challenge.

❦ Josiah Wedgwood: Beyond the Factory

For this part of the story of the Industrial Revolution, we can look at the life of Josiah Wedgwood (1730–95), whose pottery works has become one of the most iconic examples of the transformation of industry during the Industrial Revolution. Wedgwood was born into a pottery family and went into business with an older brother before starting his own company at the Ivy Works. What Wedgwood realized was that each step in the production of pottery could be separated, so that instead of a single person preparing the clay, shaping it and then glazing and finishing it, each operation could be done by someone who only did that one job. This was not an original idea, but Wedgwood formed his pottery works around this principle of the division of labor.

THE " BLACK WORKS," ETRURIA.

Figure 7.5 Etruria pottery works. Wedgwood's advanced pottery factory. Note canal passing by factory for easy transportation.

In 1769, he built a new factory named Etruria that utilized the most advanced factory techniques of the day (Figure 7.5). Constant experiments led him to systems to determine the temperature of kilns and as a result he had a far greater control over the quality of the pottery being fired. He also introduced three new types of ceramics: Queen's Ware (1762), Black Basalt (1768) and Jasper (1774).

Queen's Ware was named for Queen Charlotte after Wedgwood secured a commission to produce a tea and coffee service for her. He received permission to call his ceramic line Queen's Ware from Charlotte along with the right to say that he was "Potter to Her Majesty." This was one of the earliest examples of a celebrity endorsement, and it was a brilliant idea. The very best artists worked for Wedgwood, and their finest work was highly prized by the wealthy and the nobility. Catherine the Great of Russia ordered two different dinner sets, the larger consisting of 952 pieces. Yet the mass production methods pioneered by Wedgwood meant that the middle class could purchase ceramics that were of the same style. His products sold to almost every level of society, and were part of the increasing interest being expressed by manufacturers in marketing their products.

Wedgwood was also a leading figure in transforming the marketplace in other ways. Roads in the eighteenth century were generally bad, with most being little better than dirt tracks, dusty in dry weather and impassable in wet. In many places, the best season for road travel was winter, when the ground was frozen. Many people recognized that poor transportation was having a bad effect on business. Wedgwood and other industrialists financed the building of canals to deal with the problem of transporting large quantities of goods from the industrial regions to market and facilitate the moving of raw materials from ports and producing regions of the country. Wedgwood invested in and lobbied for the creation of the Trent and Mersey Canal, between the two rivers, and then used the canal to transport materials and finished goods from his factory that he built beside the canal. By the middle of the nineteenth century, there were more than 6,000 km (4,000 miles) of canals. The efficiency of canals was significant. A pack horse could carry 100 kg (240 lb), while the same horse pulling a canal barrage could move more than 50,000 kg (about 55 tons).

Canals were a good solution, but they had significant limitations. Foremost, they could only be built where there was adequate water. Locks could help overcome differences in elevation, but they slowed transport and were expensive to build and maintain. Following a valley was easy, but crossing higher ground was a serious engineering challenge. One of the engineering wonders of the world was the Pontcysyllte Aqueduct that carries the Llangollen Canal across the River Dee in Wales (Figure 7.6); 38 m (126 ft) high and 307 m (1,008 ft) long, it took ten years to plan and build, opening in 1805. Canals were good for point-to-point transport, but they were not very flexible when it came to adding branches or junction points.

Figure 7.6 Pontcysyllte aqueduct over the River Dee. The aqueduct was built for barge transport and was an engineering marvel.

🦿 Steam Power

While the textile industry and the concept of the factory were transforming the method of production, the invention of the steam engine would open the way for a great wave of mechanization. Steam devices go back to at least the time of the Greeks, when Hero of Alexandria described a spinning ball driven by steam. Leonardo da Vinci proposed using steam to create a cannon (the "Architonnerre") and credited Archimedes with the idea. Although Da Vinci's device was theoretically possible, it would have been slow, cumbersome and expensive, and was really beyond the technical ability of the period.

Using steam to move a piston was much more practical than trying to propel a cannonball. A number of people built steam engines in the eighteenth century, but the most important were the Newcomen atmospheric engines. Thomas Newcomen (1664–1729) was an ironmonger (a supplier of metal and hardware) and a Baptist pastor. These machines were built to help pump water out of coal and tin mines. Newcomen based his ideas on earlier work done by Thomas Savery (c. 1650–1715), English inventor and engineer, and Denis Papin (1647–c. 1713), French physicist and friend of Robert Boyle. All the early engines were very energy inefficient, but they demonstrated that steam could be turned into mechanical work. The Newcomen engines consisted of a water reservoir, a heating chamber, and a cylinder with a piston attached to a lever arm. Moving the lever arm operated the pump. The engines were called atmospheric engines because they worked not by the expansion of steam to push the piston, but rather by the condensation of steam that resulted in a lower than atmospheric pressure inside the cylinder. This was done by injecting cold water into the cylinder, rapidly cooling the chamber and reducing the pressure. The external air pressure forced the piston down until the pressure equalized and the lever arm returned to its starting position.

When James Watt (1736–1819) began looking at the steam engine, he realized that one of the major problems was the lack of precision in the components, especially the fit between the piston and the cylinder. Watt started his professional career as an instrument maker, having trained in London, but he was prevented from setting up shop in Glasgow by a local guild, the Glasgow Guild of Hammermen (whose members were "men who wielded the hammer" including blacksmiths, goldsmiths, clockmakers, armorers and locksmiths) despite the fact that there were no mathematical instrument makers in Glasgow. A group of professors at the University of Glasgow, including the important chemist Joseph Black, offered Watt work and space at the university, which was outside the jurisdiction of the guild. The opportunity proved a good one for Watt as it provided him connections with the scientific community as well as work. Watt began his work on steam engines in 1762 at the suggestion of John Robinson. His early models did not work very well, but he discovered that the university owned a model Newcomen engine, and by working on it he determined the basic principles. More important, he recognized the major problem with the Newcomen engine, which was

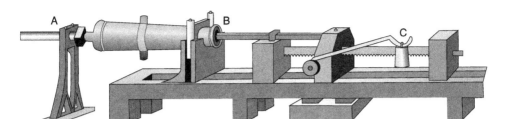

Figure 7.7 Wilkinson cannon boring machine. The cannon rotates (A) as the drill bit (B) is pressed horizontally into the metal. The lever and weight (C) maintain a constant pressure on the drill bit by forcing the bit forward with cogs along the notched plate.

the enormous waste of energy associated with cooling the piston cylinder. His solution was to build a separate condenser and thereby keep the cylinder hot.

Even with this important invention, it was a struggle to develop the steam engine into a usable commercial tool because building such a precision machine was very expensive and at the limit of the machining and technical skills of workers of the day. Watt formed a partnership with John Roebuck (1718–94), the owner of the Carron Iron Works, Stirlingshire, and with his backing created the first full-size engine. Despite receiving a patent for the steam engine, Watt was not in a position to promote his engine and Roebuck got into financial difficulty and sold his share of the patent to Matthew Boulton (1728–1809). Boulton owned the Soho Foundry, in Birmingham, and formed a new partnership with Watt that was a success. A large part of the reason that Boulton and Watt were successful was because of the work of John Wilkinson (1728–1808). Wilkinson's contribution was the ability to manufacture precision cylinders that made the steam engine much more efficient because the seal between piston and cylinder was much improved. Wilkinson had developed a method to make better cannon by casting the cannon as a solid piece and then boring out the barrel on a giant lathe (Figure 7.7). Wilkinson supplied Boulton and Watt with cylinders and also purchased steam engines for his works and encouraged Watt to expand the utility of the steam engine.

The first commercially viable Watt steam engines were installed in 1776 (Figure 7.8). The biggest market, as for the earlier steam engines, was for pumps, but as the power of the steam engine became more apparent from their successes, the demand for steam power for other purposes became greater. One of the biggest engineering problems faced by the steam engine makers was the translation of linear motion into circular motion. The most direct method was a crank, but there was already a patent on the crank and Watt was unwilling to share his steam patent in exchange for using a crank to translate linear motion to circular. Watt and Boulton got around the crank patent by introducing the sun and planet gear in 1781. This used a fixed gear on the end of the piston to propel a rotating gear on the shaft of the machine. Suddenly anything a water-wheel could do could be done by a steam engine. The advantage of the steam engine was that it could be placed anywhere and that freed mass production from waterways.

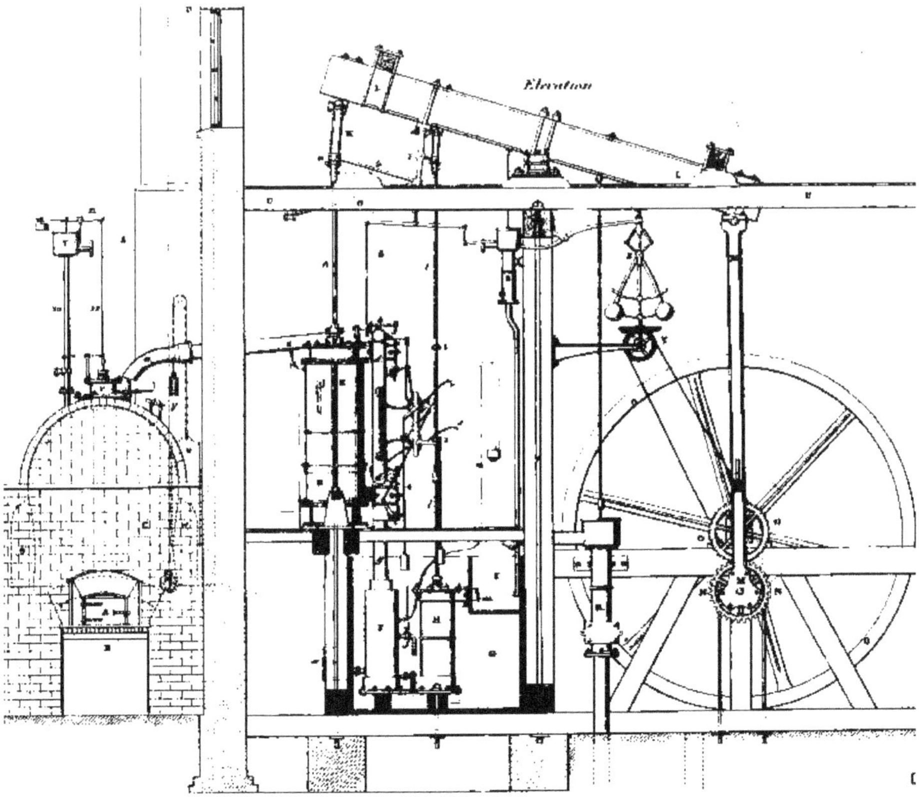

Figure 7.8 Watt steam engine. Steam passes from the boiler on the right to the cylinder in the center. The condensing steam is collected in the small cylinder below the main cylinder. Note the sun and planet gear on the large wheel and the centrifugal governor above it.

The disadvantage was cost. Waterwheels, after the initial cost of construction, were almost free, while steam engines needed a constant supply of fuel. One of the reasons that the introduction of steam power happened so quickly in Britain was that there were many coal deposits with large reserves in South Wales, the Midlands and northern England, and the Clyde and Ayrshire fields in Scotland. The steam engine created a kind of feedback loop. It increased the demand for coal while at the same time it was being used to help mine coal by powering pumps, elevators and coal trams.

✿ Steam Transport

The ability to use a piston to turn a wheel led to a transformation in transportation with the introduction of the locomotive. The earliest locomotives were little more than experimental demonstrations, but they made clear that steam could be used to propel a vehicle.

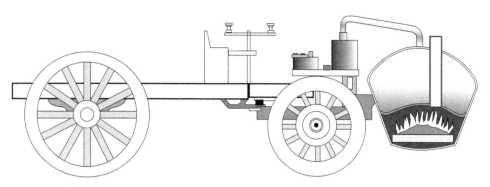

Figure 7.9 Nicolas-Joseph Cugnot's *fardier à vapeur* (steam cart) from 1769. The front is cut away to show the boiler. The cart moved, but with limited steam power and weighing over 2,000 kg (4410 lb), it was an impractical vehicle.

The evolution of the steam engine as a means of transportation followed quickly after the creation of the low-pressure stationary engine. There had been earlier attempts to create self-propelled vehicles such as Da Vinci's spring-driven cart, but the earliest steam carriage or tricycle was built by Nicolas-Joseph Cugnot (1725–1804) in 1769. It was followed by the *fardier à vapeur*, an ungainly three-wheeled monster weighing 2 tonnes with a large kettle in the front (Figure 7.9). It worked, but slowly, traveling at a top speed of about 3 km/h (2 mph). William Murdoch (1754–1839) built a working model of a steam carriage based on Watt's engine around 1784, but it did not lead to further development. Steam carriages helped to promote the idea of steam transport, but they were not practical, being too small to carry enough fuel to travel very far, as well as requiring smooth, flat roads (a rarity in the period).

⚙ Steam Locomotives

The influence of proximity to problems undoubtedly contributed to the creation of the steam rail locomotive. Mines had used tracks to guide ore cars and the early steam engines were used in the mines. Putting the two together led to a transportation revolution. On February 21, 1804 Richard Trevithick (1771–1833) drove along a tramline at the Pendydarren foundry near Merthyr Tydfil, Wales. This was followed by a number of small railways such as Matthew Murray's (1765–1826) Middleton Railway in 1812 and his engine *The Salamanca*, then *Puffing Billy* started operating in 1813 at the Wylam Colliery Railway. (*Puffing Billy* is the oldest surviving steam locomotive, on display at the Science Museum in London.) The first public railway, the Stockton and Darlington, was created by George Stephenson (1781–1848) in 1825. Stephenson is known as the "Father of Railways," and his rail gauge (the distance between the wheels) of 4 ft 8½ in (1.440 m) became the standard gauge for railways around the world.

Table 7.2 Rail mileage in the United Kingdom and the United States.

Year	UK	USA
1830	100	40
1840	4,000	2,800
1850	6,000	8,500
1860	9,000	28,900
1870	15,500	49,100
1880	27,500	87,800
1890	38,600	163,500

In the United States, the Baltimore and Ohio Railroad Company was founded in 1827 and became one of the largest and most successful. In the next decades, railways in the United State exploded, and helped expand interior trade and colonization as the tracks were built into the interior of the continent (Table 7.2). There was also the effect of the growth of railways on the speed of industrialization, as the demands for metal for rails and locomotives, the increasing power of the steam engines, and the greater engineering skill needed to create high-pressure engines for greater efficiency and power led to precision machining. Once the skills and tools of large-scale precision machining were available they could be used for a host of other industrial uses.

❧ Steel

The growth of steam power and the industrial capacity it made possible also led to a growth in mining and metallurgy. In particular, the demand for steel increased dramatically during the Industrial Revolution and changes in production techniques made it possible to mass produce steel in such quantities and at a low enough price that it became economical to use steel rather than materials such as iron, stone or wood. Steel making goes back to antiquity, with a small number of steel objects dating back to 1800 BCE. Larger-scale production of Noric steel (from a Celtic region in modern-day Austria and Slovenia) during Roman times, and of wootz steel from India and Sri Lanka, depended on a naturally occurring iron ore alloy and the invention of crucible smelting using high-temperature kilns, and ceramic vessels that could withstand those temperatures and absorbed impurities. Even at the height of ancient steel making, it was an expensive luxury item.

Figure 7.10 Blast furnace from the mid-eighteenth century. The central chamber holds the ore and is heated from below, while air is forced in from the side. The molten metal is released through a flow channel at the bottom of the furnace.

Over time, the skills of steel making degraded in Europe, so that during the Middle Ages it ceased to be available, with the exception of some steel being produced in Scandinavia, likely because the technique was brought to the region from the Middle East through the Volga trade. When the Crusaders faced Islamic forces starting in the eleventh century, they took on an opponent armed with superior weapons. There was a long fascination in Europe with Damascus steel that spurred trade and the development of metallurgy. By the sixteenth century, Georgius Agricola (1494–1555) would write about the steel manufacturing industry in his book *De Re Metallica* (1556).

At the heart of the problem of turning iron into steel was the issue of temperature. It was difficult and expensive to heat iron to its melting point and hold it at that temperature long enough to purify it and cast it. In most places, charcoal had been used for metal work, heating and cooking, and the demand led to deforestation. Coal was better than wood or charcoal for heating iron, and it became the fuel of choice wherever it was available. Useful as coal was, it had some limitations. It could vary in quality enormously, ranging from lignite (also known as brown coal) that was like a form of hardened peat and was full of impurities, to anthracite – a hard, glossy black coal with low levels of impurities, making it suitable for cooking and domestic heating. The solution to the problem was coke. Coke was coal that had been heated without oxygen to drive off water, coal-tar (a mixture of organic compounds such as phenols and aromatic hydrocarbons) and other impurities. It had been invented in China by the fourth century, probably by people who applied the idea of charcoal making to coal. It is unclear whether Europeans knew about the process from their contacts with Asia, but the first patent for "cooking" coal was granted in 1590 to Bishop John Thornborough, the Dean of York. Thornborough's interest in coal likely came from his alchemical studies.

By 1709, Abraham Darby I (1678–1717) had built a blast furnace to produce iron, using coke rather than charcoal or coal (Figure 7.10). His cast iron was used for pots and as base stock for other ironworks, but he died unexpectedly at the age of 38, leaving his work unfinished.

A new crucible method for smelting iron was invented by Benjamin Huntsman (1704–76) who discovered that using coke rather than charcoal could produce temperatures of 1600° C, high enough to melt iron and allow mixing to form the alloy steel. Huntsman was originally a clockmaker who wanted to make better steel for springs. His method worked well, but he could not find a local market and for many years sold most of his steel to French cutlery manufacturers. He did not apply for a patent, and his method was eventually discovered by other manufacturers. Huntsman set up his plant in Sheffield, and because of iron works and crucible steel Sheffield became one of the biggest industrial centers in Europe.

Steel production rose significantly because of the crucible method, but it was still largely a specialty item for tools, cutlery and weapons. Mass production of steel was made possible by Henry Bessemer (1813–98), who created the Bessemer process (Figures 7.11 and 7.12). Bessemer first describe his system in 1856 in the paper "The Manufacture of Iron without Fuel." His initial project was high-quality steel for cannons, but the demand for steel for construction and railway use was spurred on by a number of bridge disasters where cast iron structural elements collapsed.

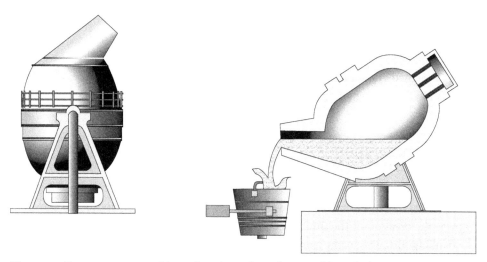

Figure 7.11 Bessemer converter. Air was forced into the molten metal from the bottom. Furnace men watched the color of the flame at the top to determine when it was ready.

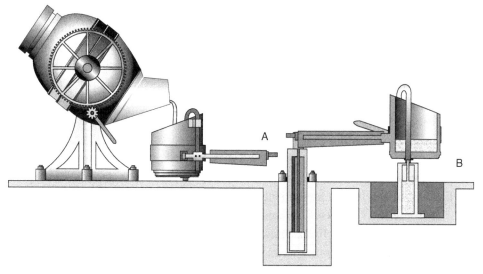

Figure 7.12 Bessemer converter with casting. The molten steel was poured into a bucket or ladle that pivoted on a hydraulic lift. It was then poured out of the bottom of the ladle into molds, leaving any slag behind.

The basic principle of the Bessemer process was to blow air up through the molten iron to burn off impurities or cause them to rise to the surface as slag that could be removed manually. The idea of using air to make iron did not originate with Bessemer, but his design and construction of the Bessemer converter made it possible to produce far more steel at a much lower price. The converter was essentially a huge crucible on pivots, that could hold up to 30 tonnes of molten iron. When Bessemer began to manufacture steel at Sheffield around 1855, he undercut other steel producers by $20 a ton, and it was not long afterwards that he began to license his system to other manufactures. Bessemer always thought big. His iron and steel mills were enormous, he designed giant stamping machines to shape iron and steel in single pieces, and he invented a way to mass produce plate glass.

The Corporation

Important as the wave of invention was, there were also two key technological developments that did not involve new machines, but facilitated the industrial age. The first was the creation of the limited company, and the second was the creation of modern banking. Britain was the leader in both of these invisible technologies, and that helped the British become a colonial power.

The concept of an organization to attend to running a business was very old, having roots in the medieval guilds and even earlier. The *commenda* or business contract had been developed during the late Middle Ages, and as the number of contracts grew, so did the number of people who specialized in business law. As these methods of organizing business became more important for trade, the money and power of the merchants steadily increased. This caused social problems, since in the social hierarchy the merchants were near the bottom but increasingly had the most money, while the upper classes were largely prohibited by tradition from engaging in commerce, but they could influence or directly control the laws under which business operated. The solution was to create organizations that took care of the business, but were not operated by the upper class. This also made it easier to have fractional investments, where many people put money into a venture. As trade in the Atlantic increased, many of the voyages were financed by this type of investment.

As trade and commerce became more significant, governments in Europe created chartered companies. In exchange for a government-granted monopoly on some product, service or resource extraction, the charter company offered important people the chance to make a great deal of money. In an era when the concept of conflict of interest was much different than it is today, the distinction between a person's role in government or the court and their responsibility to a financial venture were often nonexistent. This was one of the factors in the transformation of Europe's economy, from its earlier system based on barter and trade and an early form of capitalism known as mercantilism. Monopolies could be for almost anything, from playing cards to glass bottles, but the most significant charter companies were those that operated long-distance trade, such as the Dutch East India Company or the Hudson's Bay Company.

✿ The Great Exhibition of 1851

The greatest symbol of the industrial era was the Great Exhibition of the Works of Industry of all Nations that was held in London from May 1 to October 15, 1851. It was, in part, the British response to the highly successful Industrial Exposition of 1844 held in Paris. The Exhibition was organized by a group of leading members of the Royal Society for the Encouragement of Arts, Manufactures and Commerce, including Henry Cole (high civil servant and industrial designer), Francis Fuller (surveyor) and Charles Dilke (publisher). They gained the interest of Prince Albert, husband and Royal Consort of Queen Victoria. In 1849, Prince Albert promoted the Exhibition by stating that it would "give us a true test and a living picture of the point of development at which the

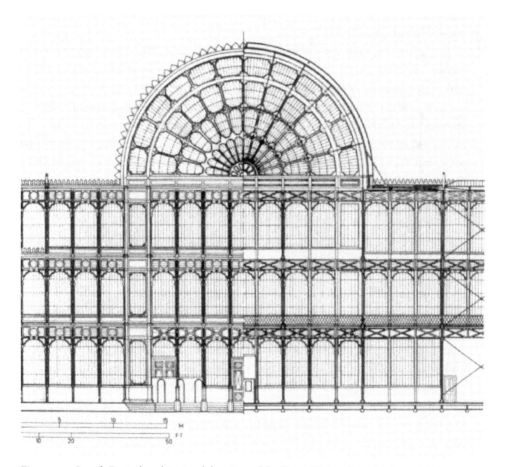

Figure 7.13 Joseph Paxton's architectural drawings of the Crystal Palace from the Great Exhibition, 1851. The building was based on modular design ideas using as many common parts as possible. Most of the components were manufactured in factories and assembled at the site.

whole of mankind has arrived in this great task of applied sciences and a new starting point from which all nations will be able to direct their further exertions."[6] With his support they raised money from the government, business and the public amounting to £200,000. They also secured commitments from important people and companies for participation. The Exhibition would have more than 13,000 displays from around the world, including a working Jacquard loom, steam engines and locomotives, kitchen appliances, arts and crafts, a reaping machine, and steel making models. The Exhibition also had lectures, examples of exotic plants, and, under tight security, the Koh-i-Noor diamond, the largest of the age. It hosted the first international photography competition, and had a public telegraph link with Edinburgh.

More than 6 million people visited the site, and the Exhibition turned a profit of £186,000 (about $36.5 million today). The money was used to fund the Victoria and Albert Museum, the Science Museum and the Natural History Museum, all of which are still in operation today.

The Exhibition building was itself a marvel of the industrial age. Designed by Joseph Paxton and nicknamed the Crystal Palace, the building used a cast iron framework and plate glass, a system that would be used for the construction of skyscrapers in the next century (Figure 7.13). It used modular construction using mass-produced and standardized parts, with most of the component parts being fabricated off-site and then assembled in Hyde Park. The building itself was 564 m (1,851 ft) long by 139 m (456 ft) wide, and 33 m (108 ft) high. It had 92,000 m² (990,000 ft²) of display space. The glass for the building covered 84,000 m² (900,000 ft²). Unfortunately, the Crystal Palace was destroyed by fire in 1936.

⚙ Energy and Chemistry

As industrialization proceeded, energy became more important. Wood had been the main fuel for millennia, but as Europe's population and industries grew, the use of wood for fuel was leading to deforestation and shortages of construction timber. Coal and coke were more efficient than wood or charcoal for heating and industrial processes, and Britain was fortunate to have large supplies of coal. As the industrial demand for coke increased, the producers discovered a serious problem: coke production created gases that could explode and a gooey, tar-like residue that had to be cleaned out of the kilns or it would ruin them. The engineer and inventor William Murdoch (1754–1839), who worked with Boulton and Watt on steam engines, experimented with coal gas and, after finding ways to purify it, used it to light his house in Birmingham in 1792. Coke manufacturers recognized the potential of coal gas, primarily a mix of hydrogen, methane and carbon monoxide; it was collected and used for lighting and heating. The first commercial use of coal gas, or "town gas" as it came to be called, was by the London and Westminster Gas Light and Coke Company in 1812. Coal gas manufacturing would remain important until it was replaced by natural gas in the second half of the twentieth century.

Coal-tar, on the other hand, was largely regarded as a waste product and dumped, often in local streams or rivers, causing ecological damage that is still a problem in places to the present day. But coal-tar, considered a nuisance industrial waste, would help change the course of the industrial world. The reason was organic chemistry. Chemistry is often overlooked as an area of technology, but those countries that developed advanced chemical industries would come to dominate international trade and be at the leading edge of technology development. Although chemical industries such as acid production had been around since the origins of civilization, they were largely craftwork, based on recipes passed down through generations. The chemical revolution started in large part by Antoine Lavoisier (1743–94) in the late eighteenth century offered the possibility of a clear understanding of chemical behavior. In scientific terms, organic chemistry deals with carbon-based compounds, so coal and coal-tar were natural objects of research. The first steps toward the development of organic chemistry for industry were made by researchers hoping to replicate organic compounds.

In 1856 the leading chemist August Wilhelm von Hofmann at the Royal College of Chemistry in London set one of his students to work trying to synthesize quinine. Quinine was the only known treatment for malaria, and its source, cinchona bark, could only be acquired from a few places such as Peru and Java at very high prices. Since the British had large colonial holdings in regions where malaria was endemic, the value of quinine was huge. The student was William Henry Perkin (1838–1907) and he used coal-tar as the source for the chemicals for his synthesis. The effort failed, but Perkin noticed that one of the compounds that he made stained cloth an intense mauve color. He had created the first synthetic aniline dye. This came at a fortuitous time. The dyes used by the textile industries around the world were all from natural sources: the plant woad produced a blue dye, crimson dye came from the cochineal insect, and purple from the mucus of the *Bolinus brandaris* snail. These dyes were never the same from batch to batch and often faded. Aniline dyes were extremely color-fast and consistent. In addition to the utility of the aniline dyes to industry, Perkin was also lucky, since Queen Victoria appeared at the Royal Exhibition of 1862 in a mauve silk gown (not dyed with aniline), making mauve the most fashionable colour for several years and assuring Perkin a financial success.

Important as the creation of the dye industry was, leading to the growth of such industrial giants as Bayer and BASF (*Badische Anilin- und Soda-Fabrik*), it was more significant as the first successful linking of modern scientific research and industrial application. Perkin opened the door for a huge range of new industrial products including plastics, pesticides, solvents, lubricants, pharmaceuticals, cosmetics, paints and dyes, and explosives. Much of the knowledge gained by studying coal-tar was also applicable to the growing petroleum industry, which would replace coal-tar as the base material for industrial organic chemistry in the twentieth century. The dream of transforming scientific knowledge into useful products had been around for generations, and indeed the Royal Society of London, founded in 1662, had as one of its objectives the creation of knowledge that would improve manufacturing, but the actual cases of intellectual transfer between scientists and industry before Perkin were limited. As the nineteenth century closed, the power of science was increasingly supported by industrialization, as leading industrialists such as Wedgwood and Watt began to fund research, starting to make significant contributions to industry. One of the most important intersections of science and technology was the discovery of thermodynamics. The new scientific ideas came in part from the observations of steam engines and the fact that cannon-boring equipment (the same tools that had made Watt's steam engines possible) produced heat continuously. In 1824, Sadi Carnot (1796–1832) would publish *Reflections on the Motive Power of Fire*, reforming our understanding of heat and energy.

As industries grew and owners looked for new products and new ways to make existing products, they began to see efficiency and greater control of the physical process of manufacturing as desirable. Scientific research offered the possibility of greater control and so science and industrialization began to replace the older, craft-based systems that had characterized manufacturing before the Industrial Revolution.

In social terms, the Industrial Revolution created not just new products, but new groups of people and social relations. Industrialists demanded economic and political power and helped overturn the remnants of the feudal system of privileges based on the old class system. The new classes of technicians (engineers, builders and mechanics) and industrial bureaucrats (lawyers, accountants and managers) grew and also gained power as the middle class – those people between the workers and the owners of industry. They were not as wealthy as the landowners and industrialists, but their level of education and collective buying power made them both necessary to the new economy and an important market for manufactured goods. Workers were at the bottom of the economic system and life was hard, but in the long run manufacturing was more lucrative and reliable employment than subsistence farming and offered greater opportunities.

The new economy, located primarily in western Europe and the mid-Atlantic states in North America, would give the people in those regions a huge advantage over people in other parts of the world. Although there had been many instances when the difference in technological prowess of civilizations was evident (the Chinese meeting almost everyone during the voyages of Zheng He, or the Crusaders meeting Islamic forces in Palestine), Western industrialization created such a vast gulf between the industrial and the non-industrial regions and peoples that we are still feeling the repercussions today.

The Meaning of the Industrial Revolution

The Industrial Revolution was not planned but people in the period knew that they were living through a period of astonishing change. The earliest known use of the term "industrial revolution" was by Louis-Guillaume Otto, a French diplomat who lived through the Terror and the rule of Napoleon. In 1799, he described France as being in competition with other nations to create new industries. The French economist Adolphe Blanqui wrote about the effects of *la révolution industrielle*, noting in 1827 that English workers were better housed, that products were less expensive and that France lagged behind in the production of machines in industry. By the 1830s the term was being used across Europe. In 1844, Friedrich Engels wrote about the *industrielle Revolution* in his book *The Condition of the Working Class in England*, and by 1881 the English historian Arnold Toynbee was lecturing about the industrial age.

Historians to this day continue to argue about why industrialization happened, if the period should be called a revolution, and what conditions led to the economic, industrial and social changes. On the pro-Revolution side is the fact that it was not just the appearance of things like factories but a vast change to the social structures of all the countries that underwent industrialization that make it a revolution. On the anti-Revolution side is the fact that it took several generations to happen and so was a more gradual process, not a sudden change, plus the fact that other events such as the French Revolution changed social relationships.

From a technological perspective, new devices were created and new systems to solve real-world problems were developed. The Industrial Revolution was characterized by three distinct technological developments:

1 mechanization of handwork
2 the replacement of muscle power with water and steam power
3 the ability to manipulate materials on an industrial scale.

These three developments led to a technological divide between those regions that industrialized and those that continued to follow traditional production methods, and that gap in power and production widened over time.

One of the other consequences of industrialization was the ability to fully exploit natural resources. Although large-scale resource extraction had been a characteristic of all powerful empires, the speed and scale of extraction grew enormously. In conjunction with the rate of extraction was the expansion of the zone of extraction, which grew from the distance a cart could move materials in a day or two to being global. This erased the economic and practical distinction between trade goods and resources. In the past, high-value trade goods such as silk and Damascus steel were often moved vast distances, but food, fuel and most raw materials had to be acquired locally. Cities and even entire empires had been established because a local resource was plentiful. In the industrial age, locality began to disappear because it did not matter if the cargo was iron ore, coal, raw cotton, frying pans, clocks, nails, books, trees or whale oil. All of it could be shipped to anywhere, from anywhere.

1 Why was it the textile industry that triggered the Industrial Revolution?
2 Was the Industrial Revolution good or bad for the poor?
3 Who were the Luddites and why did their protest against industrialization fail?

🌣 NOTES

1. The hymn "Jerusalem" ("And Did Those Feet in Ancient Time?") became England's best-known patriotic song.
2. It is highly unlikely that Wellington made this comment since Eton had no playing fields or organized team sports when Wellington attended the school.
3. Statute of Monopolies, England, 1623.
4. United States Constitution, Article I, Section 8, Clause 8.
5. The name Ludd is of uncertain origin. Some believe it comes from a Ned Ludd, or Ludlam of Anstey, who was said to have broken stocking frames around 1779. There is no reliable evidence of this, but it does suggest that anger against machines already existed.
6. Prince Albert, address at Lord Mayor's Banquet, *Illustrated London News*, October 11, 1849.

🌣 FURTHER READING

Sources for the Industrial Revolution are so numerous that any invention of the age will have dozens or hundreds of books and articles about it. As a lead in, Terry S. Reynolds' "Medieval Roots of the Industrial Revolution" (1984) links the developments of the earlier period to the growth in industry in the eighteenth century. Laura Levine Frader's *The Industrial Revolution: A History in Documents* (2006) offers a look at the period, using the words of the people themselves. A more technical look at the effect of technology can be found in Liam Brunt's "New Technology and Labour Productivity in English and French Agriculture, 1700–1850" (2002). The power and problems of empire are examined in Daniel R. Headrick's *The Tools of Empire: Technology and European Imperialism in the Nineteenth Century* (1981). One of the significant events of the Industrial Revolution was the appearance of anti-technology movements. The best known were the Luddites and a nice documentary produced by Films for the Humanities & Sciences, *The Luddites* (2007) looks at the events. To raise a historiographical question about how to understand historical classification, R. M. Hartwell's "Was There an Industrial Revolution?" (1990) provides an interesting way to think about the period.

8   The Atlantic Era

The focus of trade and industry shifted from the Mediterranean and Asia to the Atlantic region. After the American Revolution, the United States became increasingly important as a source of invention, and with most of a continent under their control, there was an abundance of natural resources to exploit. America became a place where ingenuity and inventiveness were promoted and people like Thomas Edison and Henry Ford became industrial and cultural icons. Industrialization also had a dark side, as the power of the factory was brought to the aid of the state in times of war. The First World War was the first fully industrialized war, with vast armies equipped with new weapons and giant battleships sailing the oceans. It was also the first war that involved the work of scientists directly, as chemical weapons, made possible by the growing chemical industries, were used on the battlefield.

The focus of technological change moved from the Mediterranean, especially the eastern end at the crossroads of Asia, Africa and Europe, to the western edge of Europe and the wide Atlantic Ocean. This shift was driven by a combination of greed, the search for adventure, and a desire to understand the world. What followed was the collision of European society armed with the world's most powerful weapons with the civilizations in North and South America that were as culturally sophisticated, but lacked gunpowder, naval power and a resistance to endemic European diseases. The conquest of the Americas contributed directly to the acceleration of technological development in western Europe by transforming the economy and offering various challenges that were addressed by technological innovation.

In the long run, the countries that bordered the north Atlantic became the most technologically powerful in the world, with first Britain and then the United States of America leading the world in the development and application of new technologies. The rise of the United States of America as a world power and a technological giant provides an insight into the place of technology in history. In fact, the "Yankee tinkerer" is a powerful icon of American history, representing a practical person who solves problems with ingenuity and hard work. From Benjamin Franklin to Henry Ford and Thomas Edison to the two Steves (Wosniak and Jobs) of Apple Computers, inventors are celebrated in American culture. Like the ancient Chinese age of invention, new devices and techniques have helped make the United States of America a world power.

Important as new inventions would become, interest in the "next big thing" was not a major characteristic of the early republic. The American colonies did not all rush to industrialize in the early days of the new country, but over time the demands of a growing population and access to a continent-worth of resources resulted in greater pressure to industrialize in the nineteenth and early twentieth centuries. Although the introduction of new technology sometimes had calamitous effects domestically, the early days of industrialization were in many ways a quiet revolution, since American industry was so busy supplying the domestic market that international competition really only started around the beginning of the twentieth century. American interests were then taken out to the world in a blaze of conflict, with the First World War being a major turning point in American and world history.

In 1776, the same year that James Watt produced his improved steam engine, thirteen of Britain's American colonies banded together to declare independence. Although the origins of the American Revolution are complex, one of the reasons for the Revolution was the economic inequity imposed on the colonists that was helping to fuel the growth of industrialization in Britain. Resources from the Atlantic colonies were imported at low cost to the home country, while at the same time the colonies formed a captive market for British manufactured goods. The demands by the American colonies for political representation and greater economic freedom were highlighted by protests such as the Boston Massacre on March 5, 1770 and the Boston Tea Party on December 16, 1773. Although the British government did do some things to ameliorate the situation, such as lowering the tax on tea, it responded to protests with increasingly authoritarian action. This led to warfare starting in 1775, and the war did not officially end until 1783 with the signing of the Peace of Paris, when Britain recognized American independence.

Although the conflict between Britain and her former colonies would continue to cause problems such as the War of 1812, the former colonies and the home country were tied together by the economics of the Atlantic trade. The transformation of the largely agrarian colonies from subsistence farmers to industrial power was not, however, a straightforward proposition. The power of the United States in the modern world overshadows the status the colonies on the North American continent actually had at the time of the Revolution. There was certainly money to be made in America, but initially most of the economic power on the western side of the Atlantic was to be found in the sugar islands of the Caribbean, and in South America.

Sugar has been one of the most important trade goods in human history. Although sugar can be derived from a number of sources, the majority comes from sugar cane. The original sugar cane came from Asia, with *Saccharum barberi* likely originating in India and two other species, *Saccharum edule* and *Saccharum officinarum*, coming from New Guinea. Around 350 CE, the process to crystallize sugar was discovered in India, turning what had been a kind of local snack food into a commodity. Islamic traders and farmers spread sugar cane and sugar technology across the Middle East and to the Iberian Peninsula. During the Crusades, Europeans got their first taste of refined sugar. In the twelfth century, Venetian traders operated sugar mills near Tyre (in present day Lebanon) to supply the European market.

Christopher Columbus may have taken sugar cane cuttings to the New World, but it was really the Portuguese colonists who introduced large-scale plantations and mills to Brazil and the islands of the Caribbean. By 1550, there were about 3,000 sugar mills in the region generating huge profits. The growers used native slaves at first, but found that African slaves survived the harsh conditions better, contributing to the development of the slave trade. The mills all required iron components for the presses that squeezed the sweet liquid from the sugar cane, and supplying these machined parts contributed to the growing industrialization of Europe.

The sugar islands, and in particular the island of Hispaniola (modern-day Haiti and the Dominican Republic), were one of the most important destinations for the Atlantic trade. The "triangle trade" could be worked in several ways, but formed a triangle between Europe, the west coast of Africa and destinations in the Caribbean, South America or North America. Each point provided certain commodities. Europe supplied manufactured goods ranging from textiles to firearms. Africa supplied slaves, ivory, gold and salt, while the New World markets produced sugar, rum, tobacco, coffee, furs, precious metals and cotton. In the late eighteenth century, Hispaniola supplied about 40 percent of the sugar and 60 percent of the coffee consumed in Europe. To produce these goods, one in three African slaves was sent to the island to work on the plantations. The total number of slaves taken for the Atlantic trade is not certain, but estimates indicate that between 9 and 14 million slaves arrived in the Americas, and up to 25 million people were displaced (taken as slaves to the New World or enslaved and moved in Africa). About 8–10 percent of the African slaves went to North America.

A trade voyage might take manufactured goods such as firearms, beads and copper from England to Africa in exchange for slaves. The slaves were sold to sugar plantations and sugar was taken back to England. The actual patterns of trade were more complex, depending, for example, on the political relationships among the nations trading in the Atlantic, with French, English, Spanish, Portuguese, Dutch and American interests all vying for profits, sometimes at war, sometimes allies. Traders ranged from big companies such as the Hudson's Bay Company, to small companies operating one or two ships or freebooters always looking to find the next profitable cargo, either by trade or by piracy.

The importance of the Atlantic trade led to innovations in naval technology, since the demands of long-distance trade favored faster and larger ships. Vessels capable of traveling from Europe, across the Atlantic and through the Pacific to Asia, particularly to China, had to be designed and built, leading to the age of the tall ships. The new ships such as the schooner and the clipper began to sail the Atlantic and Pacific trade routes in the middle of the nineteenth century. The schooner was gaff rigged (a four-cornered sail with a spar at the top of the sail) and, according to nautical tradition, was first built by Andrew Robinson in 1713 at Gloucester, Massachusetts. Popular as a pirate ship, it was in the industrial era a fast transport.

The first clipper was the *Annie McKim*, built for the company Kennard & Williamson at the port of Baltimore in 1833. Clippers were square rigged having a large rectangular sail supported by a long beam called a yard suspended across and at right angles to the mast. The part of the yard that extended past the sail for the control ropes was called the yardarm. The clippers were usually bigger than schooners and were used for freight, especially the "China clippers" that carried tea. These ships used the most advanced devices such as capstans, pulleys and rudders controlled by large wheels to control the large ships with a minimum number of crew. The smaller the crew, the more space available for cargo and the lower the cost of operating the ship. This helped to make the Atlantic trade very profitable, and in turn spurred the development of other industries.

One of the important (but not very glamorous) businesses that grew up because of the shipping trade was insurance. Long-distance trade was dangerous and ships were frequently lost to weather, navigational problems, bad crews or poor maintenance. In order to prevent the loss of a ship from bankrupting companies or investors, money lenders took up the practice of offering insurance. This was essentially a gamble: the insurer was betting that the ship would return safely from its voyage in exchange for earning a fee. The amount of the fee depended on what the insurer thought was the risk of loss. A long voyage by an inexperienced crew in an old ship would cost much more to insure than a short trip by an experienced crew in a well-maintained ship. One of the most famous organizations that worked in this industry was Lloyd's of London. Lloyd's was not an insurance company, but rather a market for insurers (or under-writers) that started informally at Edward Lloyd's Coffee House in 1688. In 1760, the Register Society was formed by customers of the market with the objective of providing a rating system for ships. The first published Register Book appeared in 1764 and listed ship owners, masters, home port and other information, but most crucially a rating of the condition of the hull and rigging. The insurance system pushed owners to keep up their ships, spread the risk of investing in shipping and generated profits that helped to finance further industrial development.

The introduction of new ships by American builders was an important techno-logical achievement, but responses to the technological innovations of the Industrial Revolution were very mixed in the early days of the new nation. Many of the lead-ers of the Revolution had strong philosophical leanings, and their dedication to the ideals of the Enlightenment included a profound interest in science and technology. Benjamin Franklin, along with co-founders George Washington, John Adams, Thomas Jefferson, Alexander Hamilton, Thomas Paine and others, established the American Philosophical Society in Philadelphia in 1743. The stated objective of the society was to conduct "philosophical Experiments that let Light into the Nature of things, tend to increase the Power of Man over Matter, Conveniences or pleasures of Life." In other words, explore nature for the sake of knowledge, but keep an eye open for practical applications as well. It was from such philosophical and practical interests that in 1780 John Adams founded the American Academy of Arts and Sciences in Boston, along with Franklin and Washington.

Given the interests of Franklin, Adams and the other leaders, it is not surprising that concerns about invention appear in the Constitution, which was written by many of the same people. Article I, Section 8, of the Constitution grants the federal government right and duty "To promote the Progress of Science and the useful Arts, by securing for limited Times to Authors and Inventors the exclusive Right to their respective Writings and Discoveries." The Patent Act of 1790 would clarify the degree of protection, saying that inventions that were deemed useful, important and new would be granted fourteen years of protection. By 1836, some 10,000 patents had been granted. While that might seem like evidence of a lively technological scene, in fact many of the patents were for devices that never became commercially useful or, in the case of designs for various fanciful flying machines, were physically impossible.

Despite the interest in intellectual property rights in the Constitution and the efforts by Franklin and other leaders to promote science and technology, the actual level of industrialization in the post-Revolution period was not outstanding. The country continued to be dominated by agrarian interests, with economic power largely resting with the southern states. Europe continued to be the source for much of the manufactured goods purchased by North Americans. There was also a certain reluctance to fully employ patent protection, since there was money to be made copying the designs of things imported from Europe. Even as the factory system and steam power was being introduced in English mills, North American producers stayed with water power and local production.

The Industrial Revolution in America shared with the European Industrial Revolution strong links to changes in the textile industry and the introduction of power technology, specifically steam power. In social terms, the European experience of industrialization was accompanied by social disruption, as a result of both the military events of the period, particularly the French Revolution and the rise of Napoleon, and the creation or expansion of the working and managerial classes. Although the old wealth based on land and inherited privilege did not disappear, in those countries that experienced the greatest level of industrialization there was also the greatest shift in social standing as the mill owners moved into the upper classes and the farm workers became the mill workers in the increasingly urbanized industrial regions.

The situation in the United States was somewhat different. Although there were landed interests, particularly (but not exclusively) in the southern states where large-scale plantations using slave labor operated, there was no gentry or noble class and land ownership was far more widespread. Slaves, who were used both as agrarian workers and in many of the mills and factories, were a very economical labor force, decreasing the drive to replace expensive workers with machines, a major factor in European industrialization.

As the interior of North America was colonized, European settlers displaced the native peoples and established thousands of hectares of new agricultural, timber and mining lands. As settlements multiplied and population grew, the demand for man-ufactured goods also increased. The northern states such as Pennsylvania and New York that had access to coal, iron and other resources started to industrialize to meet the market demands. They could compete with European goods in part by copying European designs with little fear of litigation, but also because they had a much shorter supply line and so could reduce transport costs.

American industrialization was closely linked to transportation. The continent was big, and the population very spread out. In the early part of colonization, population centers were located on the sea or the larger navigable rivers. When more goods and materials needed to be transported, the problem of distance and topography limited options. In England there had been a canal building boom, so it is not surprising that the first proposals to improve transportation in America were canals. The most signifi-cant canal proposal that actually got built was the Erie Canal that connected New York on the coast to Buffalo on Lake Erie and thus to the Great Lakes. This would provide mass transportation from the Atlantic trade routes to the interior of the continent. Suggestions to make such a connection went back to the late seventeenth century, but the technical difficulties were prohibitive, particularly the fact that the canal would have to climb 180 m (600 ft) from the level of the Hudson River to Lake Erie, some 580 km (360 miles) away (Figure 8.1). Large sections of the canal required little more than muscle power to dig, but the locks and aqueduct required to complete the canal were just at the edge of possible construction techniques available at the start of the nineteenth century.

The Erie Canal was promoted by Jesse Hawley, who wanted to grow grain in upstate New York for sale in the Atlantic trade, but transportation costs bankrupted his idea. With help from land agent Joseph Ellicott, Hawley convinced New York governor DeWitt Clinton and the state legislature to invest $7 million in the canal. Building started in 1817 and was completed in 1825. It was an incredible engineering feat, par-ticularly considering the fact that there were no professional civil engineers in America and the surveying and planning were done by amateurs. The canal was a financial suc-cess and reduced transportation costs to the interior by over 90 percent. There was a huge cost in lives, however. More than 1,000 construction workers died during the building of the canal, the majority overcome by disease, particularly malaria and yellow fever during the building of the section through Montezuma Marsh at Cayuga Lake. Others died from rock falls, blasting accidents or construction injuries.

On the north side of the Great Lakes, the Welland Canal would bypass the Niagara Falls in 1829, and, along with locks on the St. Lawrence River, shipping on the Great Lakes would be connected directly to the Atlantic. The St. Lawrence Seaway, com-pleted in 1959, became one of the most important transportation systems in the world.

Economical as canals could be once they were constructed, they were not an option for most of the continent. The solution to the problems of mass transportation was

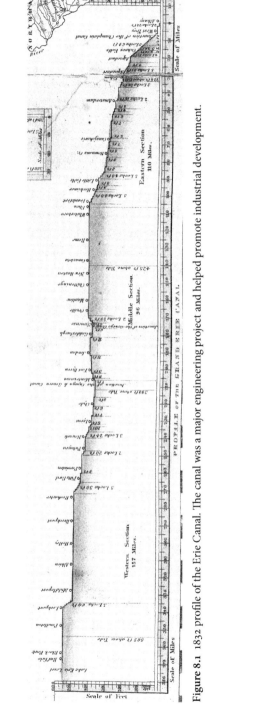

Figure 8.1 1832 profile of the Erie Canal. The canal was a major engineering project and helped promote industrial development.

railways. Although the engineering difficulties were still considerable, rails could go just about anywhere. A number of small rail-based transport systems were built in the late eighteenth and early nineteenth centuries, such as the 1,200 m (¾ mile) horse-drawn rail cart system at Thomas Leiper's quarry in Pennsylvania in the 1780s and the steam locomotive designed and built by John Stevens in Hoboken, New Jersey in 1825, but they were mostly for specific and local use. The first general carrier was the Baltimore and Ohio Railroad. After spending a year in England investigating how railways were built and operated, Philip E. Thomas and George Brown founded the B&O in 1827, with the idea of connecting the port at Baltimore, Maryland to the Ohio River. Construction started at Baltimore in 1828 and the line reached the Ohio River at Wheeling, West Virginia in 1853. A separate line connecting Baltimore to Washington DC was completed in 1835 and would be an important transportation route during the Civil War.

A railway boom developed between the 1830s and the 1860s, with dozens of companies seeking to establish rail companies in various parts of the country. In the early days of the transportation boom, canals and railways were seen as a means of connecting important economic centers to the eastern seaboard so that goods could be sold into the Atlantic trade. One of the most profitable commodities sold to the Europeans was cotton.

✿ Cotton in America

Cotton may have been planted by the Spanish in Florida as early as 1556 and small plots were planted in Virginia in 1607. Despite the potential demand for cotton, it was an expensive product because it required high levels of labor. This arose from the fact that the cotton fibers had to be pulled from the seed pod and then the seeds separated from the fibers, or "lint." Cotton rollers, a simple device consisting of a wooden or metal cylinder set so close to a platen that only the lint could pass between them, could be used to extract the seeds from black seed cotton, but the majority of cotton in America was green seed cotton and the rollers would not work. Thus, every seed had to be removed by hand, and even with slave labor, this was a long and costly job. In 1790, only 1.5 million pounds of cotton were produced in the southern states.

A mechanical solution to this problem was pursued by a number of people, including Michael Almaviva, Catherine Littlefield Green, Sean Paul and Joseph Watkins, but the first successful cotton gin (gin being a contraction of engine) was created by Eli Whitney in 1792. Whitney was granted a patent on his machine in 1794, but a combination of a weak patent law and a poor business plan meant that Whitney never made the profits he hoped to gain from his invention.

Table 8.1 Cotton production.

Year	Production in millions of pounds
1790	1.5
1800	35
1830	331
1860	2,275

The basic principle of the cotton gin was to use a hook or claw to draw the lint through a narrow opening. The opening was too small for the seed, which would pop off the fiber. A brush then removed the lint from the hook. A series of hooks mounted on a cylinder turned by a manual crank (later by water or steam power) allowed large quantities of cotton to be processed in a very short time.

The growth of the American cotton industry was explosive after the introduction of the cotton gin, and the value of cotton to the economy of both the cotton grow-ing states and the northern states that dominated the shipping industry grew just as swiftly (Table 8.1). The textile mills of England eagerly bought the cotton, and the profits fueled industrialization in the USA and England. It also led to a great increase in demand for slaves, and the number of slaves rose from around 700,000 in 1790 to almost 4 million in 1860.

Whitney's original plan had been to set up cotton gin operations around cotton coun-try and take two-fifths of the processed cotton as the fee for cleaning the cotton. He would have been fabulously wealthy if he had been able to control the cotton gin in that man-ner, but the high price and inconvenience quickly led plantation owners and local smiths to copy his device. Whitney came close to bankruptcy attempting to protect his patent. Whitney might have failed as a businessman, but he became an icon of American inven-tion. In the process, the cotton gin changed the course of American history. It created capital for investment, spurred the development of transportation routes and made the United States a more important part of the Atlantic trade. It also affected the social struc-ture of the country, especially in the cotton-growing states, and set the stage for a conflict between the agrarian south that depended on slave labor and the industrial north.

⚙ Transportation

Moving goods and people north and south shaped American culture. Railways were a means to move people and goods cheaply from place to place, and to a large degree they were successful in achieving that objective. What the creators of the rail systems, both in the New World and in Europe did not appreciate was that the railways would have a profound effect on society. Like the rivers of the great ancient empires, the railways were a communication system. Not just people and cargo traveled the lines; information also moved at a significantly faster rate. Events in distant towns and cities could be known in a few days or even a few hours rather than the weeks that it took mail to travel by horse or boat. One of the most profound aspects of the new speed of information flow was the ability to centrally manage businesses. Unlike the age of the chartered companies, when "factors" were sent out to oversee business in distant outposts, but were largely autonomous so long as they pursued the general goals of the company, business owners could now control daily activities from their head offices. Lucrative mail contracts were awarded to rail companies by the federal government both as a means of supporting the railways and to guarantee this communication system.

⚙ The Clash of Technology

As America grew into its independence, it began looking outward. Americans had been sailing to the Orient for many years, mostly looking for trade goods such as tea, silk and ivory. If a ship could survive the long journey, there was a great deal of profit to be had from importing these luxury items to North America or Europe. Yet the Oriental trade was dominated by European interests and many of the Asian countries limited trade in various ways, preventing American interests from setting up large-scale commerce. In particular, Japan had closed its ports to all foreigners except for a very limited trade with the Dutch.

In 1846, Commander James Biddle was sent with two ships by the United States government to Japan to attempt to gain a trade agreement. He had successfully negotiated a trade agreement with China in 1845, but was unsuccessful with the Tokugawa government. In part because Japan *was* such a closed market, it seemed to offer great promise to the country that could open trade with Japan. When in 1848 Captain James Glynn successfully negotiated the repatriation of American whalers who had been shipwrecked in Japan, interest in Japan was increased. Glynn, having observed the lack of modern weapons in Japan, recommended to the American government that a show of force might convince the Japanese to open trade.

On July 8, 1853, Commodore Matthew Perry sailed his four warships into Uraga Harbour near Edo. His armament outclassed anything the Japanese had and he used the threat of bombardment of the city if he was not allowed to open negotiations. On July 14, Japanese officials allowed Perry to land at Kurihama and present a letter from President Millard Fillmore to the representatives of the Tokugawa Shogunate. The letter outlined a trade agreement.

Perry then left for China, but returned in February 1854 with eight ships to receive the reply. The representatives of the Tokugawa government had accepted almost all of

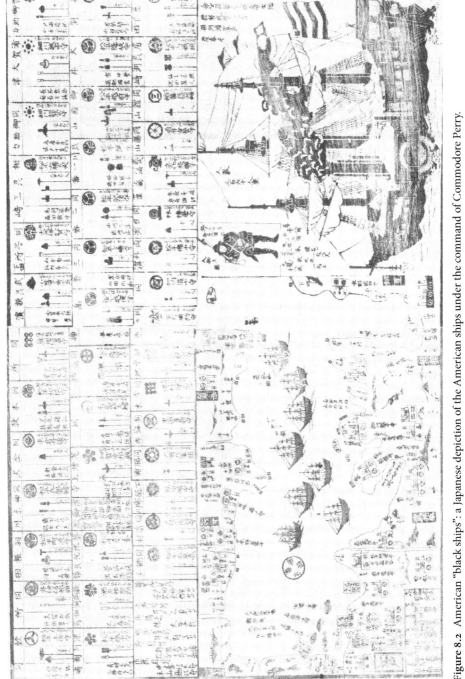

Figure 8.2 American "black ships": a Japanese depiction of the American ships under the command of Commodore Perry.

the American demands. Perry returned home to great acclaim, and was even awarded $20,000 by Congress for his work. The voyage of the "black ships," as the Japanese called the American warships, had opened Japan to trade (Figure 8.2).

The story of Perry and the black ships has often been presented as an example of the effect of technological inequality, with the Japanese surrendering to the superior weapons of the American navy, and opening the doors to economic imperialism by the West. It has also been portrayed as a blow against isolationism and the draconian methods used by Japanese officials to keep foreigners out of Japan in order to maintain a feudal system. Both of these contain elements of truth, but the story is more complex and it hinges on the evaluation by the Japanese of the technology they observed and what was happening in other parts of Asia. The internal politics of Japan were complicated, with many rival factions and a split between the imperial government that was largely powerless and the shogun, who controlled real power. Perry's trade deal offered officials in Japan a way to turn a dangerous situation to their own advantage. Perry's warships could have caused a great deal of damage, but it would have been very difficult for the Americans to actually force the Tokugawa Shogunate to open Japan to trade without a prolonged struggle. It was a calculated risk that by accepting trade, even on bad terms, Japan would gain technological equality in the future and avoid many of the problems being experienced by their Asian neighbors. In a very short time, Japan did transform itself, rising in military power to the point that it defeated China in the Sino-Japanese War 1894–95 and extending its sphere of influence over much of the western Pacific region. In effect, the Japanese created their own "black ships."

The success of America's first major exercise of power in the international realm did not lead to a more general expansion of American interests as the United States slid into the Civil War and a period of internal turmoil during the post-war Reconstruction era.

✿ The American Civil War: New Technologies of Destruction

The American Civil War was one of the bloodiest conflicts of the modern era, with some 620,000 soldiers killed and an unknown number of civilian deaths. More than 400,000 soldiers returned home wounded. The causes of the Civil War were complex, but from a technological perspective, the struggle represents a divergence of cultures and the increasing industrialization of war that had first appeared during the Napoleonic era.

The southern states were primarily agrarian and depended on slaves as a major source of labor for plantation farming, but also for resource extraction, domestic labor and factory work. The economy was based on exports such as tobacco, sugar and especially cotton for the textile mills of England. The northern states, particularly those on the Atlantic seaboard, were increasingly industrial, and as the century progressed tens of thousands of European immigrants were traveling to the northern states, filling the cities and from there pressing into the interior of the continent. While it would be a mistake to see the people in the south as racists and the people in the north as completely egalitarian, slavery was much more strongly opposed in the north, where slave culture had never been widely practiced and where slaves were seen as a threat to paid workers.

The leaders of both the north and the south were aware that their cultural and economic interests were diverging and that the opening of the interior to settlement would have a great deal to do with the continued viability of each side. If the north gained control over the vast resources and limited the use of slaves, the south would be blocked from expansion, both territorially and economically. The south already had access to the great artery of the Mississippi River that led to much of the interior, and hoped by establishing slave owners' rights in at least the southwestern region that they would gain the economic and territorial power they needed to challenge the growth of the north.

When war finally came in 1861, the south's military leaders believed, with good reason, that they had the superior soldiers. The average southern soldier was more accustomed to life in the field, was far more likely to have grown up hunting and riding, and in addition, they believed they were defending their homes and way of life since many of the battles were fought in the southern states. What the south lacked was industrial capacity and population. By the second year of the war, northern forces were bigger and their foundries and factories were vastly out-producing what could be made in the south. By the end of the war, the Confederacy had lost the capacity to manufacture heavy weapons and were unable to mass-produce even small arms. With even a partial blockade of the southern ports by the northern naval forces, materiel from Europe for the war effort was severely limited. Just as the British had used their industrial capacity to defeat Napoleon, the northern states could put more men in the field, with great numbers of guns, shells, tents and wagons. While the industrial advantage did not guarantee victory, it made it far more likely.

In addition to the mass production of battlefield equipment, the Civil War was the first war in which railways and the telegraph played a major role. In 1860 railroads were heavily concentrated in the northern states, with more than 35,405 km (22,000 miles), compared with only 13,745 km (8,541 miles) in the Confederate states. In addition to utilizing all available commercial telegraph lines, the Union forces installed 24,766 km (15,389 miles) of lines during the war. This can be compared to only a few thousand kilometers of lines in the Confederate states, primarily associated with the railways. During the course of the war, the Union Army sent over 1 million telegrams and President Lincoln sent an average of twelve telegrams a day to his commanders in the field. The centralization of information gave the Union forces a much better tactical and strategic sense of the military situation. The superior communication technology would change the way wars were fought, transferring some of the control over the battlefield away from the front-line officers to the commanders at a greater distance from the fighting. While some of the front-line control was given up, commanders could call for supplies and support in a way that made the infrastructure of war far more important. While southern forces were better able to operate independently and live off the land, that form of warfare (often becoming a form of guerilla warfare) was not effective in a war that increasingly focused on control of population centers and strategic supply lines.

✿ Total War

These applications of technology were part of the transformation of warfare from an isolated action to what became known as "total war." For most of history, wars were fought by very small numbers of combatants in very localized conditions, typically clashes in the field or sieges of fortifications. Many were won or lost in a single battle, and people only a few kilometers away might not even know there was a battle raging. Total war means that a nation at war diverts the work of the whole society to war production. Although there are examples reaching back to ancient times of societies such as the Spartans devoting themselves to war, total war was really a product of the industrial age, when industrial capacity that had been dedicated to commercial production was redirected or even retooled to make things for the war. This extended beyond factory owners attempting to fill contracts for the military to the introduction of centralized control of industrial capacity by the government and the appropriation of resources (such as minerals, food and power) as war materiel.

✿ Interchangeable Parts

One of the aspects of military industrial development associated with the Civil War was the increasing use of interchangeable parts for weapons, particularly small arms such as muskets. Various forms of standardization had been introduced to weapons such as artillery over the years. A good example of the problem of standardization comes from the Spanish Armada that attacked England in 1588. Since every cannon was individually handcrafted, the artillery officers of the Spanish fleet each carried a set of rings that represented the diameter of the cannons they were commanding. They used the rings to sort through the cannonballs available to find ones that would fit their guns. While this could be done ahead of the battle, it was a slow and often frustrating activity. If it had to be done during the course of battle, it might make the difference between winning and losing.

By the nineteenth century, the manufacturing of artillery was much more uniform, so that clearer classes of weapons such the "12-pounder" used by Napoleon's Grande Armée, or the 32-pounders carried by the British Navy were standardized enough that artillerymen didn't have to check the diameter of each ball. It was still the case, however, that weapons were generally uniform by manufacturer, but they were not compatible from maker to maker.

The idea of uniformity of parts regardless of maker comes primarily from Jean-Baptiste Vaquette de Gribeauval (1715–89), a French artillery officer. Gribeauval both introduced regulations about standardization of French artillery and advocated a system of production that would allow parts to be manufactured in different foundries and work together. This required a much greater degree of control over designs and the introduction of specifications based on precise measurement rather than individually crafted production tools. In other words, the major change was not that the musket parts had to be the same, but that the machine tools used to manufacture the musket parts had to be the same, and the manufacturers had to read and understand the design specifications in the same way.

❧ Photography

In addition to the changing technology on the battlefield, the American Civil War was one of the first major conflicts that was extensively photographed. Because of photography, we have a much clearer historical record of the conditions of war than from any previous conflict. Photographers such as Mathew Brady, George S. Cook and Alexander Gardner actually went out to the battlefields and traveled with military units. The images were created using daguerreotypes, or the more flexible calotype, or talbotype process. In 1827, Joseph Nicéphore Niépce (1765–1833) produced the first photograph, but it was Louis-Jacques-Mandé Daguerre (1787–1851) who in 1839 discovered a manageable method for capturing images on glass plates. The patent for the daguerreotype was acquired by the French government, but rather than monopolize the process, the government announced that the system was given as a gift to the world. Although tens of thousands of daguerreotype images were made, Daguerre's system was largely overshadowed by the system created by William Henry Fox Talbot (1800–77) in 1834 and publicly displayed in 1839. The talbotype, or, as it became better known, the calotype, produced prints on paper. Talbot fought to protect his patent, but when innovations were made he did not pursue a patent renewal after his original patent expired in 1855. By 1860, photography had become a commercial activity with companies manufacturing cameras, chemicals and printing equipment.

❧ The New Europe

While America began the long and painful process of rebuilding after the war, in Europe the consequences of industrialization on international politics were beginning to play themselves out. With access to a market of hundreds of millions of people and the natural resources from around the globe, Britain had created the most powerful

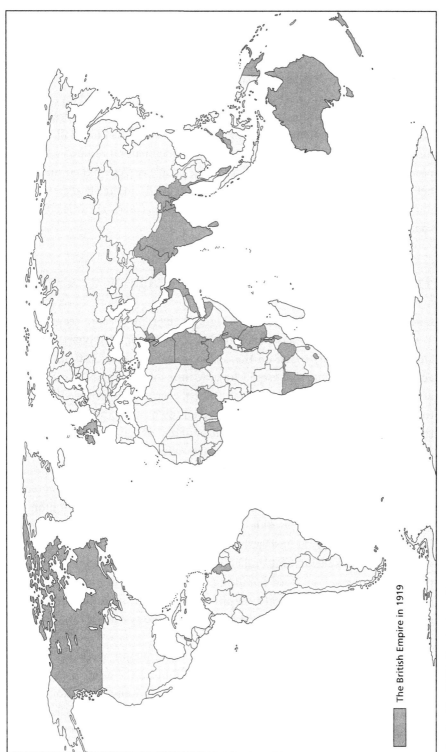

The British Empire in 1919

Map 8.1 The British Empire in 1918.

empire in history (Map 8.1). For a variety of reasons, but in part as a response to British power, there were two major unifications. First Italy, with the exception of the Papal City, unified the Italian states in 1861 under Victor Emmanuel of Sardinia. In 1870, the Germanic states, except Austria, became a united Germany under Kaiser Wilhelm and the "Iron Chancellor" Otto von Bismarck. This was the era of the "Great Game" as the European powers sought to gain land and power through politics, war and trade. Yet underlying the actions of all the European powers, great or small, were the growing demands of industrialization. Industries needed natural resources and the nations needed the products of industry to supply the people and equip their militaries. Falling behind was psychologically bad, but also potentially militarily disastrous. This was made quite clear during the Franco-Prussian War (1870–71) when German forces soundly defeated French forces.

✿ Time and Place in the New Industrial Age

One of the problems created by the telegraph, the railways and steamships was time. Traditionally, local time was established by local custom and observation. The town clock was set, usually based on the astronomical observation (with wildly varying accuracy) of noon, and that established the time for the region. Since noon occurred at a different time for every spot on the globe, each city, town and village had its own time. When trains and telegraphy connected towns and cities, the problem of local time being different everywhere became apparent as schedules and planning were a mess or impossible to create. In Britain, the railways began to use Greenwich Mean Time (GMT) in 1847, and by 1855 most of Britain was using standard time. In North America, a consortium of railways in the United States and Canada agreed to adopt standard time and five time zones for the continent starting at noon on November 18, 1883. This was only possible because the Allegheny Observatory in Pittsburg could telegraph a signal at exactly 12 o'clock across the country so that everyone would know the specific time and adjust their clocks. This helped standardize time in North America. Sandford Fleming (1827–1915), inventor and railway engineer, proposed worldwide standard time in 1879, although it would take until 1929 for most of the world to accept Universal Time and the system of time zones on which it was based.

Time zones and standard time can only work if people agree on how to measure where they are on the globe. Within the British Empire, the prime meridian (the line of longitude set at zero degrees) was set at Greenwich, the home of the Royal Observatory. In order to coordinate navigation and timekeeping, US President Chester A. Arthur convened the International Meridian Conference in 1884. It established Greenwich as the prime meridian, although some countries such as France resisted the use of a

British location for mapping and setting time. By 1911 the prime meridian, standard time with twenty-four time zones and the International Date Line had become the norm for navigation and timekeeping and were kept accurate by a network of observatories around the globe.

To this day, the responsibility of establishing time is an important, if often hidden, responsibility of national governments. Time is important for transportation and scheduling, but it is also crucial for everything, from legal events such as the specific moment real estate changes hands, to stock transactions.

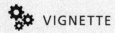 VIGNETTE

Charles Goodyear: A Cautionary Tale

Inventors in the golden age of American amateur invention from 1800 to 1900 have become icons in American history. People like Samuel Morse and Thomas Edison are remembered as men whose inventions reshaped the world and made money from their work. Charles Goodyear (1800–60), who was a contemporary of Morse, has also become an icon of invention, but his story has been used as a cautionary tale (Figure 8.3). Goodyear's invention of the vulcanization of rubber stands as a tale of trial and error, the recognition of a business opportunity, and a lack of judgment that turned the business opportunity into something of a tragedy.

In the first part of the nineteenth century, natural rubber was being used to manufacture a variety of objects including tubing and life preservers. The problem was that rubber became a sticky mess in hot weather, turned hard and cracked in cold temperatures, and even in mild conditions tended to rot and fall apart. Goodyear started to experiment with additives and processing methods in 1832, producing some promising results but never quite getting the rubber to set perfectly. His family came near to starvation while he continued his experiments. Finally, in 1838 he discovered that the addition of sulfur and high heat would "cure" the rubber and transform it into a firm, temperature resistant but still resilient material. There has been much speculation about how this discovery came about. Goodyear claimed it was trial and error in his tiny attic laboratory, but another story about spilling the rubber/sulfur mix on a hot stove comes from the same period. In either case, Goodyear recognized that he had solved the technical problem, but not the business problem. He, along with two brothers, set up a factory in Springfield, Massachusetts, but did not have the backing to start large-scale production. He made terrible royalty agreements and spent large amounts of money defending his patents in America.

Goodyear also sent samples of his rubber to British manufacturers hoping to interest them in licensing his process. British inventor Thomas Hancock (1786–1865) of Charles Macintosh and Company, saw some of Goodyear's rubber and that seems to have steered him to the process of vulcanization, named after the god Vulcan by one of Hancock's friends. Hancock filed his patent before Goodyear filed in Britain. Hancock had been working for several years on stabilizing rubber before getting the samples and it is unlikely he could have figured out the process simply from the samples, so his patent represented actual research, even if it was based on Goodyear's work. Goodyear sued Hancock and was offered a half-share of Hancock's patent to drop the suit, but refused and ultimately lost the legal battle. He lost his French patent as well, and even though he was awarded the Cross of the Légion d'Honneur by Napoleon III for his invention, he still ended up in a debtors' prison. Perhaps because of the years of poverty and exposure to toxic chemicals, Goodyear's health was not good, and he collapsed and died at the age of 59 with outstanding debts of $200,000. In 1898, Frank Seiberling commemorated Charles by naming his company the Goodyear Tire and Rubber Company, although no one from Charles Goodyear family was ever employed by the company.

Figure 8.3 Charles Goodyear.

⚙ Engineers as the Agents of Invention

While Goodyear represented the solitary inventor working for years in relative isolation, the other approach to technology creation was engineering. Engineers, especially what we would call today "civil engineers" (people who design, build and maintain physical structures such as buildings, roads and bridges), have existed since the beginning of civilization, but it was only in the eighteenth century that people began to go to school to learn about engineering. The first civil engineering school was the École Royale des Ponts et Chaussées (Royal College of Bridges and Roads), founded in Paris in 1747. "Engineering" was not an academic discipline at the time, so the students learned mathematics, particularly geometry, basic physics and hydraulics. They visited building sites and worked with the builders and designers in something that combined academic studies and apprenticeship. The École Centrale des Travaux Publics (Central School of Public Works) was founded in 1794, and was renamed the École Polytechnique a year later. It emphasized mathematical and science studies, but always had a strong emphasis on applied science and many of its graduates became engineers. The École Polytechnique survived the many revolutions and wars in France and continues to be one of the leading engineering schools in the world today.

In the United States, the earliest institution to offer engineering education was West Point (United States Military Academy), founded in 1802 and modeled in part on the École Polytechnique. The title of oldest engineering school is somewhat disputed, with Norwich University teaching civil engineering as early as 1819, while the first degree in civil engineering was granted by Rensselaer Polytechnic Institute in 1835. The University of Virginia School of Engineering and Applied Science, established in 1836, was the first American school of engineering. The first doctorate in engineering was awarded by Yale University in 1863, while the most famous engineering school, the Massachusetts Institute of Technology, was founded in 1861 and opened its doors to students in 1865. It was to be a school dedicated to "industrial science" according to its charter of incorporation.

In Britain, there had been attempts to teach engineering at the beginning of the nineteenth century, but they failed. The first engineering department that lasted more than a few years was at King's College London, started in 1838, followed by Glasgow University in 1840 and University College London in 1841. In India, the Thomason College of Civil Engineering was founded in 1847 and would later become the Indian Institute of Technology after India gained its independence.

In Germany, technical schools (*Technische Hochschule*) started to appear in the eighteenth century, with the oldest being the Collegium Carolinum founded in 1745 and becoming the Technische Universität Carolo-Wilhelmina zu Braunschweig (shortened in English to Brunswick Institute of Technology). One of the most famous is the Technische Universität Bergakademie Freiberg (Freiberg University of Mining and Technology), founded in 1765 and the oldest continually operating university of mining and metallurgy in the world. Its motto is "The University of Resources" and the elements indium and germanium were discovered by researchers at the university.

It is not surprising that formal education in engineering came at the same moment that the Industrial Revolution was helping to create a mass society. Long-distance travel, growing urbanization and increased population meant that the traditional way of finding an expert to work on an engineering project was breaking down. Direct contact or even personal recommendations were not sufficient in the growing industrial world. Although experience and recommendations would (and still do) play a part in the employment of engineers, educational credentials became a reliable way of determining who had the skills to take on the growing number of projects made possible by industrialization. Over time, the number of engineering schools would grow, as the complexity of industrial society grew. Engineering schools would also start offering specialized training, starting with civil engineering and mechanical engineering and adding electrical and chemical engineering around the beginning of the twentieth century. The training of technological specialists would contribute to an acceleration in invention, and was increasingly required to maintain existing technology. By the end of the nineteenth century, the development of infrastructure such as roads, railways, water systems, gas lines and the telegraph were benefiting from technical innovations that were only possible with specific and specialized knowledge of the systems themselves. Individual inventors were still important, but teams were becoming more important as the members of a team could provide a wider range of technical knowledge.

In turn, the networks of engineers with similar education would work to create standards for engineering covering everything from building codes to the classification of tools. This helped to protect the personal interests of the engineers, who would become both the inspectors and the producers of standardized materials, but it also improved the safety and reliability of the material goods, buildings and devices of everyday life. There was also a darker side to the education of engineers. From at least the time of Archimedes to the halls of West Point and the École Polytechnique (which is still supervised by the French Ministry of Defence), engineers have always been associated with war.

⚙ Technology and the Path to the Great War

In 1885, the Berlin Conference was held. It brought together delegations from the European powers to discuss trade relations and to divide up the last remaining colonial territory in Africa. Germany, growing more powerful and having recently defeated France, was hoping to gain useful territory, but the Germans came away from the conference disappointed. Minor powers such as Belgium and Portugal controlled more valuable African territory than Germany, while the French and British owned or controlled areas such as South Africa and Egypt that were economically and strategically important. Although there were historical roots to the colonial holdings, and Germany was a latecomer to the colonial game, its lack of access to natural resources from colonies was a key factor in shaping German domestic and foreign policy.

In order to overcome its deficits, the German government, in cooperation with business and the educational system, created a powerful science and technology system. The heart of this was education. German policy was to educate everyone, and by 1900 the general level of education in Germany was the highest in the world. The brightest boys (and it was almost exclusively boys) were identified at the elementary level and streamed toward higher education at the universities, while those with mechanical or technical inclinations were sent to technical schools to learn trades. At the universities, the most scholarly students in the liberal arts would become the next generation of professors and teachers, while the others went on to jobs in business and government. Those interested in technical subjects became engineers, some staying to teach at the universities, and the others going to work for industry and government. In science, there was an additional layer, since the government and private business interests created a group of research institutes called the Kaiser Wilhelm Institutes. The top researchers went to work at the KWIs, the next rank were researchers and professors at the universities, and the next tier went to work in industry.

This created both a formal network and a vast informal network, so that the interests of industry could influence the areas of research, and researchers could communicate their discoveries to people who might be interested in real-world applications.

Germany needed to feed its growing population, but it could not draw on colonial holdings as Britain could, or vast tracts of arable land as France or Russia had within their borders. The only solution was to get more food from the land that Germany controlled. To do that, the German government encouraged the development of intensive agriculture, introducing mechanization, and training agronomists and farmers in the latest techniques. Important as the move to industrial-scale agriculture was, it required one thing that Germany had limited access to, namely nitrates for fertilizer. In the nineteenth century, the greatest source of nitrates came from islands on the west coast of South America that were covered in hundreds of meters of guano. These deposits were so valuable that they contributed to the War of the Pacific (1879–84) that saw Chile fighting Bolivia and Peru. Bolivia lost the territory of Litoral, leaving itself landlocked. Most of the nitrate production in South America was controlled by British companies. The other major source of nitrates was India, and was therefore also under the control of the British.

To solve this problem, Germany turned to its scientists. The lack of nitrates for agriculture and industry was one of the best examples of the German system at work. The chemist Fritz Haber, working at the University of Karlsruhe, in 1908 discovered a process to "fix" atmospheric nitrogen. In other words, he came up with a method to cheaply turn nitrogen, which makes up about 78 percent of the atmosphere, into ammonia (NO_3). Ammonia could in turn be converted into fertilizer, industrial chemicals and explosives. In 1909, Karl Boch (a graduate of the Königlich Technische Hochschule), in cooperation with the steel company Krupp, created the specialized steel vessels and equipment to turn Haber's discovery into an industrial process. One of the most significant chemical discoveries was industrialized in less than a year, freeing Germany from dependence on natural nitrates imported from the Pacific. It also provided Germany with the base stock for the explosives it would use during the First World War.

♯ Electricity

The history of electricity extends back to antiquity. The Greeks knew about what we would call static electricity, Aristotle speculated about the nature of lightning and some archeologists have even suggested that the Egyptians could have made primitive batteries. The problem with electricity was that it was so difficult to control. To study it required the ability to consistently create it and store it until it was needed, but this proved to be difficult. The invention of the Leyden jar, consisting of a glass bottle wrapped in metal with a metal rod through a cork in the opening, opened the door for study. The Leyden jar was a capacitor, holding a charge that could be generated by a static electricity source. The charge could then be released. This offered the ability to put electricity where an experimenter wanted, but it discharged completely and that limited what kinds of experiments could be done with it. Benjamin Franklin used a Leyden jar to do experiments concerning lightning, particularly using lightning rods to protect buildings from damage.

Alessandro Volta (1745–1827) discovered that there was a link between chemical activity and electricity. He placed thin plates of metal, zinc and copper separated by paper soaked in an acid solution, and created a battery that provided a constant flow of electricity. The voltaic pile made electricity much more manageable, and extended its use beyond the laboratory (Figure 8.4).

In 1819, Hans Christian Oersted (1777–1851) was preparing to do a public demonstration of various physical principles, including the flow of current and magnetism. He noticed that the needle of a compass moved when current ran through a nearby wire. This was the first discovery of the link between electricity and magnetism. André-Marie Ampère (1775–1836), following Oersted, worked out many of the principles of

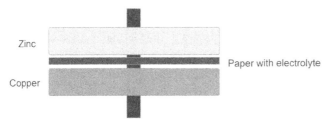

Figure 8.4 Voltaic pile. The pile or battery allowed for the controlled use of electricity. More electricity could be generated by adding more layers.

electromagnetism by 1827. In the same year, Georg Simon Ohm (1789–1854) worked out what would become Ohm's law, that the current of electricity is equal to the ratio of the voltage to the resistance in the circuit.

In 1829, Joseph Henry (1797–1878) began his work with electromagnets, demonstrating that a core of iron would be strongly magnetized if it was wrapped with a coil of wire carrying the electrical current. During his work, he also discovered the principle of the dynamo, but Michael Faraday (1791–1867), who was also working on electromagnetism, published first. A dynamo generated electricity by moving a wire through the magnetic field of a permanent magnet. This would be the foundation of electrical generation. In 1831, Henry designed an electrical motor, in which a current passing through coils caused an axle to spin.

❧ Telegraphy

With so much more known about how electricity worked, plus the increasing ability to produce and control electricity, people began to think of it as something that could be used outside the physics laboratory. The idea of electrical communication was widespread. Francis Ronalds (1788–1873) built a small telegraph using static charges in 1816 and described its use in his book *Descriptions of an Electrical Telegraph and of Some Other Electrical Apparatus* (1823). Ronalds anticipated telegraphic systems, suggesting wires across England and writing: "let us have a small steam engine, to work a sufficient number of plates to charge batteries or reservoirs of such capacity as will charge the wire *as suddenly* as it may be discharged when the telegraph is at work" (Ronalds 1823: 96).

Baron Pavel L'vovitch Schilling (1786–1837) working in St. Petersburg, Russia used static charges and galvanometers to transmit electrical signals up to 5 km (3 miles). It used electrical transmission to generate a magnetic field to move an arrow to point at the letters of the alphabet, spelling out words. The system was cumbersome and slow, but its utility was obvious. Carl Gauss and Wilhelm Weber operated a telegraph in Göttingen, Germany in 1833. In 1837, Charles Wheatstone and William Cooke patented a multi-wire telegraph and built a version that ran 2.4 km (1.5 miles) from Euston to Camden Town. In 1838, the Great Western Railway had a Cooke and Wheatstone telegraph installed between Paddington station and West Drayton. It used a series of five needles linking six wires that powered magnets that moved the needles either to the right or left to encode messages. The first commercial telegraph was started by Wheatstone and Cooke in 1839 in conjunction with the Great Western Railway.

Telegraphy was ideal for controlling train traffic, and commercial telegraphy initially grew out of the control systems for railways and used the rail right of ways for the wires. Because sending signals could be slow and expensive, there were great efforts to come up with code systems to compress and speed up transmission. The most famous system was developed by Samuel Morse (1791–1872) and his assistant Alfred Vail (1807–59). They introduced Morse code in 1838 and the Morse key, a simple but elegant device to send the signals. At almost the same time as Cooke and Wheatstone were introducing their telegraph, in the United States, Morse and Vail created their simpler two-wire electromagnetic telegraph and developed the Morse code system of dots and dashes.

By 1850 a telegraph cable was laid between England and France. By 1861, telegraph wires connected the east and west coasts of the United States. After several attempts, in 1866 the Atlantic Telegraph Company linked England and the United States, and a cable from Britain to India was completed in 1870. Suddenly, communications that had taken days or weeks to travel by sea or rail could be completed in minutes. Mass communication by telegraph meant that news from across the country and eventually around the world could inform people about events beyond their immediate neighborhood while at the same time fostering a sense of nationalism and unity. It would also change the way wars were fought by centralizing command.

✿ Commercial Electricity

The second problem was more technical. Power generation initially produced direct current (DC) which provided electricity at a constant voltage and electron flow traveling in one direction. This was fine, since the devices invented to use electricity such as incandescent lights and motors worked perfectly with DC current. The problem was that DC could not be sent very far over power lines. This meant that power had to be produced close to the place where it would be consumed. In turn, this meant there had to be many power plants, and that was not very economical. The solution to the transmission problem was alternating current (AC) that changed both the voltage and the direction of electron flow. AC could be transmitted over longer distances, thus making it economical to have large power generators. A number of engineers and inventors attempted to solve the transmission problem using AC. In 1884 the Hungarian engineers Károly Zipernowsky (1853–1942), Ottó Bláthy (1860–1939) and Miksa Déri (1854–1938) created some of the first commercially successful AC generators, transformers and electricity meters, while in Britain, Sebastian de Ferranti (1864–1930) developed AC generators and transformers that were installed in the Grosvenor Gallery power station for the London Electric Supply Corporation in 1886. At the forefront of AC technology was Nikola Tesla (1856–1943), whose patents for generators, motors and other electrical equipment were licensed by George Westinghouse (1846–1914) in the United States.

The struggle over the commercial distribution of power using AC or DC pitted the two great American power promoters against each other, with Westinghouse favoring AC and Thomas Edison supporting DC. The "War of the Currents," as the struggle was sometimes called, was as much a publicity campaign as a battle of technical specifications. Edison and General Electric tried to associate AC with danger to users and fatal accidents, and to demonstrate this publicly killed animals with AC. Edison hired lobbyists to try and influence state legislatures to outlaw AC power systems, and even secretly funded the invention of the electric chair as a means of execution to emphasize its danger. In 1890, William Kemmler was the first person to be executed by electric chair, at Auburn Prison, New York.

A series of power company mergers resulted in there being only two large electrical companies: General Electric (which took control of the Edison Electric Light Company) and Westinghouse Electric Company. Despite the negative campaign by Edison and General Electric, Westinghouse won the war, because AC could be far more easily distributed. A major turning point came in 1893 when Westinghouse won the right to construct a power plant at Niagara Falls. The plant would produce AC using Tesla's devices. The contract had been awarded on the recommendations of an international commission headed by the famous scientist Sir William Thomson (later Lord Kelvin) (1824–1907). AC power was supplied to the important industries in and around Buffalo. General Electric was not completely dealt out of the power system, since it had quietly been adding AC equipment to its own system and was awarded the contract to build the transmission lines. Eventually, power from Niagara Falls would supply a significant part of the electricity used on the eastern seaboard in the United States and Ontario. Edison, who had lost control of his electrical company in 1892, retired from the electricity business and turned to other projects.

Although AC became the standard form of power delivered to residential and commercial users (heavy industries often used DC power from their own power plants), local domestic use of DC power continued in a few places in the United States and Europe as late as 2007. This was made possible by the use of rectifiers that converted powerline AC to DC at or near the consumer. In an interesting twist of technological development, the actual use of DC has grown enormously in recent years as modern devices such as cell phones, laptop computers, cordless tools and halogen lights use converters to change AC current to DC. Essentially, any device that uses a battery is built to use DC.

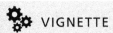 VIGNETTE

Thomas Alva Edison: Epitome of American Inventors

Edison was born in Milan, Ohio and grew up in Port Huron, Michigan (Figure 8.5). He was the last of seven children, and received little formal education. His life was changed when he saved a child from being struck by a runaway train, and in gratitude the boy's father trained Edison as a telegraph operator. Naturally curious, his introduction to telegraphy sparked a life-long interest in technology, particularly communication technology. He invented a number of devices to do with telegraphy before he made a success with the phonograph. His interest in sound technology may have been spurred by his own poor hearing, the result of scarlet fever and ear infections (not, as is often told, being hit in the ear by an angry train conductor). After selling his quadruplex telegraph to Western Union for $10,000 in 1874, he set up a research laboratory at Menlo Park, New Jersey in 1876. It was the first industrial research center in the United States and one of the few in the world.

At Menlo Park, Edison brought together a team of researchers and created or improved dozens of devices. Most notable was his work on the electric light bulb. Edison did not invent the light bulb but worked with a team, and nor was his team the first to make an incandescent light: there were more than twenty incandescent lamps created before his team began to work on the problem. The problem was not the light bulb itself, it was making a light bulb that was reliable and inexpensive to manufacture. What Edison was a master at was conceiving inventions as part of systems and rationalizing the component parts as production objects. Thus, Edison understood that the light bulb was only another addition to the electrical system, and the Edison Company became the producer of the first commercially successful line of light bulbs for residential and commercial use.

Figure 8.5 Thomas Edison.

In many ways, the creation of electricity as a commercial product is comparable to the discovery of the controlled use of fire. It opened the way to a massive wave of invention and marked the beginning of a new form of technology that worked autonomously. Although pre-electrical mechanical clocks, automatons and some carillons had independent control mechanisms, electricity removed the necessity of human intervention in the operation of many devices. So long as the current was running, electrical devices would work autonomously. An electric clock, even though the mechanism was still a system of gears, never needed winding. A refrigerator never needed a new supply of ice blocks, and electric lights needed no wicks, fuel or matches to work. Early electrical devices often used the older mechanical control systems (an electric motor rather than weights in a gear-driven clock, for example) but as electrical devices became more complicated, inventors did more than create electrified forms of existing devices, they created a new order of artifacts: devices whose function was controlled by the electrical state of component parts. Devices could have timers, sensors for a variety of triggers such as temperature, pressure, moisture or sound, and later radio waves and light.

In addition to the autonomous capability of the new machines, they also created a new kind of technological barrier for the user. They were the first devices where an observer could no longer determine the method of operation of the device without specialized knowledge. The age of electricity created, as a necessary corollary, the age of the technician.

Ford and the New Factory System

While Edison was helping to electrify the world, Henry Ford (1863–1947) was beginning his transformation of the transportation industry. Ford was responsible for many innovations, both great and small, but he is often given credit for things he didn't really invent. For example, he did not invent the factory system, nor did he invent the internal combustion engine. Although he was a pioneer in the degree of the division of labor in manufacturing, Wedgwood and others had introduced the division of labor generations before Ford. It was two innovations, one well publicized and one less well known, that made Ford so important in the history of technology. The well-known innovation was the automated assembly line. Rather than moving the workers and materials to the assembly area, Ford moved the automobiles and the parts to the workers. This required a much greater control over the component parts, and Ford's early automobile designs were based on the minimum number of parts possible. These innovations meant that the time needed to produce an automobile dropped from 12 hours to 90 minutes. The Model T Ford was the first Ford automobile to be built using the new techniques and it first rolled off the assembly line in 1908. It transformed transportation. Rival companies, first in the United States and then around the world, would adopt his methods for manufacturing everything.

Ford's other innovation was the pursuit of vertical integration of the manufacturing process. Again, this was not an idea that originated with Ford, but he set new standards for large corporations to follow. This meant Ford tried to own everything from the iron mines, the smelters, the parts manufacturers, the assembly plants, the delivery system and the showrooms. By owning each part of the network, he could control costs, delivery schedules and standards of production. Although Ford was not completely successful in controlling all aspects of the industry (his efforts to get into petroleum did not get very far), he differed from most of his contemporary industrialists because he saw the automobile as part of an industrial system, not just as the product to be sold.

As other companies adopted his production methods, the market for automobiles grew and so did the competition for market share. One way to keep the public interested in buying new cars was to introduce new models. Ford was initially opposed to changing designs, which he saw as wasteful, but the success of the Model A starting in 1927 and the pressure of competition led to new models annually. While most of the changes are cosmetic (tail fins and the shape of windows), other innovations such as the electric starter (1912), hydraulic brakes (1920), automatic transmissions (1934), disk brakes (1949) and power steering (1951) make driving easier and safer.

✿ Petroleum World

Making automobiles helped to transform transportation and had broad effects, but the industry would not have been possible without petroleum. In the early days of automobiles, steam-powered and electric cars competed for the consumer market, but the internal combustion engine using refined petroleum won. One factor in this was simply the choice of the car makers to go with petroleum fuel, which in turn spurred research and development. The most significant reason, however, was range per unit of fuel. Electric cars were perfect for the urban environment where trips were short and electricity available, but long distances were a problem. Steam engines needed both fuel and water and could not travel as far with the limited amount of fuel that could be conveniently carried.

Yet at the beginning of the twentieth century it was not clear that the petroleum industry could supply enough fuel for the growing transport market. Crude oil had been known since ancient times, with surface deposits in places like the La Brea tar pits in California or the Binagadi asphalt lake in Azerbaijan being used by local people for things like waterproofing boats and as a source of flammable material. In the ninth century, the Persian alchemist al-Razi experimented with crude petroleum and discovered a clear liquid that burned with no smoke. This substance would be rediscovered in 1846 by Abraham Gesner (1797–1864), who called it "kerosene." Gesner used coal as the source for the fuel, and distilling kerosene from coal was a multi-step process and that made it expensive. In the 1850s, a number of people began to experiment with petroleum to replace coal. Samuel Martin Kier (1813–74) in the United States and partners Ignacy Łukasiewicz (1822–82) and Jan Zeh (1817–97) in Poland began to market both kerosene and lamps to burn it. Because their kerosene was produced from petroleum rather than coal, the earlier patents were circumvented. Kerosene eventually replaced whale oil and other products as an illuminating fuel.

The demand for kerosene meant there was a market for petroleum and discoveries were made of oil deposits in Pennsylvania, Ontario and southern Poland. As more kerosene became available it was used for heating and cooking, which in turn increased demand. Kerosene was a major commodity at the beginning of the twentieth century and further prospecting led to the discovery of major deposits in Texas, Venezuela and Russia and large deposits in the Middle East, particularly Saudi Arabia.

The refining of kerosene produced other substances such as tar that was used for asphalt, but most of the other by-products were discarded. When the gasoline fraction was used as a fuel for the internal combustion engine, it created a secondary market for petroleum. With the growth of the automotive industry, gasoline and later diesel fuel became more important than kerosene. The introduction of electric lighting further decreased the demand for kerosene, although it is still manufactured for traditional oil lamps and for use in aviation fuel.

By the middle of the twentieth century, a vast network of production, refining, transport and distribution of fuel for transportation and heating had been built up. Other products, including lubricants, plastics and organic chemical bases, were important but secondary markets for petroleum. The majority of oil deposits were owned or controlled by European and American companies, but the international situation started to change in 1960 with the formation of the Organization of the Petroleum Exporting Countries, or OPEC. The initial five countries (Iran, Iraq, Kuwait, Saudi Arabia and Venezuela) grew to fourteen members by 2016. OPEC members represent about 44 percent of world production and hold an estimated 74 percent of the known oil deposits.

Arms Races and the Nature of Industrialized War

One of the political by-products of industrialization was the increasing realization that the control of resources was imperative for national interests and could not be left in private hands. As the power of Europe and America grew, the political leaders and many people in the general population felt that only by direct confrontation could the interests of the state be kept secure. What followed was an arms race that introduced more powerful versions of existing weapons and new types of weapons to war. The weapons tell us two important things about the application of technology to war. First, and most obvious, is that the weapons represented a significant increase in the damage that could be done compared to the weapons used in the nineteenth century (Table 8.2). This was partly based on the weapons themselves, with high explosives replacing black powder and bolt action rifles and machine guns replacing muskets, but the real

Table 8.2 Comparison of artillery rounds expended (Crowell and Wilson 1921: 27).

Battle	Date	Force	Rounds	Rounds per day
Battle of Gettysburg	1863 3 days	Union Army	32,781	10,927
Battle of Gravelotte	1870 1 day	German Confederation	39,000	39,000
Battle of the Somme	1916 7 days	British Expeditionary Force	4,000,000	571,429

source of destruction was the scale of weapon use. Industrialization and the assembly line made everything on the battlefield available in quantities never seen before. The primary example of this can be seen with artillery, where more rounds were fired at the Battle of the Somme than were used in the entire American Civil War and the Franco-Prussian War combined.

The second and less obvious lesson the weapons tell us is about management. The arms race and the industrialization of war required a growing managerial capacity in the military, industry and government that would allow vast projects such as the building of dreadnoughts to be successfully undertaken. Technical specialists such as engineers, electricians, metallurgists and chemists could invent new things, but bringing them together in the complex systems of production also required factory managers, military commanders, accountants, government bureaucrats and lawyers.

❧ Dreadnoughts

Two crucial technological developments changed navy construction at the end of the nineteenth century. They were the introduction of steam-powered ships and the replacement of wood with iron and steel construction. One of the leading pioneers of the new ships was the famous engineer Isambard Kingdom Brunel (1806–59). In 1859, he launched the *SS Great Eastern*, a massive 210 m (700 ft) steamship designed to carry 4,000 passengers from England to Australia without stopping for fuel. Although the *Great Eastern* failed as a passenger ship, she was used to lay transatlantic telegraph cable. It took more than twenty years for ship builders to fully integrate the innovations that Brunel brought to naval construction, but by 1900 battleships had become technological marvels. They used the latest forms of steam turbines, electrical communications and lighting, and mechanical systems for everything from loading the guns to moving cargo.

Figure 8.6 *HMS Dreadnought*, 1906. The biggest battleship of its time, it set off an arms race when other countries rushed to build their own superships.

One of the most notable products of the new industrial power of Germany and America was their efforts to challenge British naval power by building larger and larger ships and fleets. The pinnacle of naval design and a major spur to the arms race was the launching by Britain of the *HMS Dreadnought* in 1906 (Figure 8.6). The *Dreadnought* was massive: 160 m (526 ft) in length, weighing 22,200 tonnes fully loaded, and armed with ten 12-inch guns, twenty-four 12-pounders and five torpedo tubes. She was more than 30 m (100 ft) longer than the next largest Nelson-class ships. She cost a record-setting £1,785,000 ($211,967,000 today). The construction of the dreadnoughts was part military and part international relations. Since it was impossible to keep construction on that scale secret, the British government's commitment to building the massive new ships was seen as sending a warning to other countries about British naval power. Germany, France and other countries were not to be outdone, and began their own dreadnought program (Table 8.3).

Impressive as the dreadnoughts were, they played almost no part in the outcome of the war. The Battle of Jutland, the one major sea battle between British and German forces, was indecisive. While the Royal Navy lost fourteen ships compared to Germany's eleven, Germany had fewer ships and could not withstand that level of attrition. *HMS*

Table 8.3 Dreadnoughts or dreadnought-scale ships produced by 1916 (Moore 2001).

Country	Number of ships	Country	Number of ships
Britain	28 dreadnoughts 9 dreadnought battlecruisers	Austro-Hungary	4 dreadnoughts
Germany	16 dreadnoughts 5 dreadnought battlecruisers	Russia	4 dreadnoughts
USA	10 dreadnoughts	Argentina	3 dreadnoughts
France	4 dreadnoughts 3 super-dreadnoughts	Chile	3 dreadnoughts
Italy	6 dreadnoughts	Brazil	3 dreadnoughts
Japan	5 dreadnoughts	Turkey	2 dreadnoughts

Dreadnought was outdated technology by 1914, spending most of the war guarding the British coast. She sank only one enemy vessel, ramming and sinking a German submarine in the Firth of Forth. The arms race in dreadnoughts was one of the best examples of the dilemma of the "Dollar Auction." Fear of losing a naval confrontation led countries to build more of these ships than could be of any practical use. Commanders were reluctant to expose their dreadnoughts to actual combat, especially when it became clear that dreadnoughts could be damaged or sunk by inexpensive mines, U-boats and much smaller ships armed with torpedoes.

❧ Submarines and Sonar

In contrast, submarine warfare proved, at least at the beginning of the war, to be extremely effective. In 1914, Germany had twenty-nine U-boats, and they sank five British cruisers in the first ten weeks of the war. By the end of the war, Germany had put 360 U-boats in the water, and over the course of the war U-boats would sink more than 10 million tons of shipping. Britain, France and Russia also used submarines, but they played a more limited role in the war. In response to submarine attacks, new naval tactics were developed and detection equipment built. Early work on sonar (detection by underwater sound waves) by Reginald Fessenden (1866–1932) in 1912 and Paul Langevin (1872–1946) in 1915 led to the development of the ASDIC sonar system developed for the British Admiralty.[1] It was a different kind of arms race between attackers and detectors. Submarines were still a danger, but sonar gave surface ships a chance to fight back.

⚙ Machine Guns

Another of the inventions that changed the nature of war was the Maxim machine gun and its myriad offspring. Hiram Stevens Maxim (1840–1916) was born in the United States and emigrated to England. He was not the first person to work on repeating-fire weapons, the most famous early weapon being the Gatling gun introduced in 1861 in time for the American Civil War. What made Maxim's guns different was that they used the recoil of one shot to reset the firing mechanism and load the next cartridge, making the Maxim gun completely automatic after the first shell was fired. Maxim said of his invention that he met an American in Vienna in 1882 who told him: "If you want to make your everlasting fortune and pile up gold by the ton, invent a killing machine – something that will enable these Europeans to cut each other's throats with greater facility – that is what they want" (Maxim 1914: SM8).

Maxim's gun was not an immediate commercial success, in part because the machine gun did not fit well into the existing categories of weapons. Although it was far smaller than any artillery piece, it was not a weapon that could be carried or used easily by a single person. The doctrines of the British and the French armies were still largely based on massed infantry forces armed with bolt-action rifles and bayonets. The French put some of their machine guns with the artillery because they were capable of indirect fire, or shooting at targets that were not in line of sight. The Germans were the first to recognize that machine guns worked best as an infantry weapon, and the speed and lethal fire of the gun would allow a smaller force to face larger numbers of soldiers.

The opening of the First World War saw the Germans win some early encounters with the French, encouraging them to think they could repeat their victory of the Franco-Prussian War. The machine gun made short work of French infantry assaults and cavalry charges. The fighting, however, quickly bogged down, with neither side able to bring enough force to overwhelm the opposing side in a direct assault. The German and Allied sides both attempted to get around the flank of the enemy, but since both sides were attempting the same maneuver, all this did was spread the front line out over a longer and longer distance.

⚙ Chemical Warfare

Germany had entered the war believing that its well-trained and technically sophisticated military could defeat France, but with the help of the British the technical advantage was either removed as both sides adopted similar weapons and tactics, or nullified by the greater resources that the British could throw into the fight. For example, successful as the German U-boats were, they could not defeat the British naval block-ade or end the supply of resources that flowed into Britain. In a war of attrition, the Germans were in a weaker position. The German military turned to chemical war-fare. Germany had a well-established system of scientific research through the Kaiser Wilhelm Institutes. Fritz Haber, director of the Kaiser Wilhelm Institute for Chemistry and Electrochemistry, and the chemist who had discovered how to synthesize ammo-nia out of atmospheric nitrogen for fertilizer and explosives, suggested using chemicals as a weapon.

The German Army launched its first major chemical attack on the Western Front on April 22, 1915. They released chlorine gas from 5,700 cylinders that had been labori-ously moved to the front lines. A greenish-yellow cloud blew on the breeze across the French position, and because chlorine is heavier than atmospheric air, it settled into the trenches and dugouts. Chlorine was an easy choice. Large supplies were available because it was a prime material for dozens of industrial processes, from cloth bleaching to water treatment. Chlorine injures or kills by combining with moisture in the mouth, nose and lungs to form hydrochloric acid. It was a painful death. The French troops, unprepared for the surprise attack, retreated, leaving many dead and injured behind. A gap more than 6 km (4 miles) wide opened and the German Army pushed through, but because the attack had been planned as a trial, there were not enough troops to force a major breakthrough.

What followed was a chemical arms race that saw more and more toxic chemicals reach the battlefield. All sides geared up for chemical warfare, building or expropriating chemical factories for war work, and at the same time working to defend soldiers, pri-marily using gas masks. The most toxic materials to be used were phosgene and mus-tard gas. Mustard gas was not really a gas at all, but an oily mist that could coat things and remained toxic in the environment for days or even weeks.

Chemical weapons were partly designed to kill the enemy, but their bigger pur-pose was to be a psychological weapon and to overwhelm the support system, especially hospitals, behind the lines. Although the Germans were world leaders in the chemical industries, they were not so far ahead of the French and English that chemicals changed the course of the war. It simply made trench warfare that much more deadly for all.

An important by-product of the introduction of chemical warfare was the recogni-tion that science and scientists were also a form of national resource. Because of chem-ical warfare, most of the major combatants set up government-funded research. Many of those organizations created to bring scientists together to fight chemical warfare, such as the National Research Councils of the United States, Canada and Australia, continue to fund scientific research.

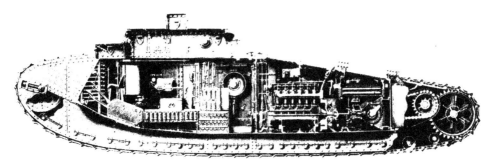

Figure 8.7 British Mark VII tank. The introduction of the tank helped to end the stalemate of trench warfare.

⚙ Tanks

Armored vehicles had been used since ancient times, but it was the invention of the internal combustion engine that made mechanized weapons possible. The first use of tanks was by the British on September 15, 1916 during the Battle of the Somme, but the first real success come on November 20, 1917 when about 400 tanks of the British Tank Corps led an advance of almost 9 km (6 miles) at Cambrai. The Germans attempted to build tanks in response, but by 1917 they were running into supply problems and only managed to build a handful. These clanking monsters, weighing as much as 43 tonnes, were designed more like moving bunkers than assault vehicles, but they helped turn the war in favor of the Allied forces. The tank was a good example of invention as an amalgamation of existing devices. It combined the engine of the motorcar with machine guns and cannons, all wrapped in the armor plate typical of the battleship (Figure 8.7).

⚙ Aviation

The final notable innovation of the war was aviation. Although airplanes have received a great deal of romantic attention, particularly about aces like the Canadian Billy Bishop (officially credited with seventy-two victories) or the German Red Baron Manfred von Richthofen (eighty confirmed victories), they had very little direct effect on the war. While air combat received the most popular attention, reconnaissance both on land and at sea was more significant to the war effort. What the war did do was act as a kind of beta test, leading to ideas such as strategic bombing, ground support, aerial photography and fighters. It also convinced a number of young officers that airpower would be vital for future wars.

✿ The World after the War

At the end of the war, the balance of world power had shifted. Britain and Germany, the industrial leaders before the war, had been surpassed by the United States. Although the USA been producing more materials like steel than its European counterparts since the turn of the century, it was the war that really brought America into the international arena. The American interest in invention, a growing economy and the resources of most of a continent gave the Unites States a huge boost.

It was not Yankee invention that was the greatest contribution to the development of technology in the Unites State; it was the creation of a new society. The Republic, with its founding documents enshrining Enlightenment ideals of social equality and freedom, revised almost every aspect of previous forms of government and social organization. The founding fathers, well versed in the classical history of the Greek polis and the Roman Republic, appreciating the power of the British parliament and learning from the native confederations, strove to create a new organization. The government was so radical that many, even among those who supported the revolution, did not really believe it would work. There was, for example, a movement to make George Washington monarch of the new country. He declined. In the early days of the Republic, presidents wondered if they could rally enough support to keep the office viable. The power of the Supreme Court was at best theoretical for many years, and it was not unusual for elected representatives to Congress to fail to show up or to leave during sessions as their personal business dictated.

Despite the early problems, the great power of the new country was its dedication to social, intellectual and personal freedom. This is not to say that such freedoms were completely available to all, as the long struggle over slavery shows, but compared to the rest of the world, the United States offered possibilities that nowhere else could offer. This drew people from across the globe. The invisible technology of American government had been designed with the ideal of bettering society, not merely controlling it. In part, they succeeded because they had the tools to make the dream of an Enlightenment society a reality. The founders had also, partly by intent and partly by luck, created a society that was not just open to novelty and innovation, but fostered it. Even today, American inventors and technological innovators are better known than military leaders or politicians.

1 Why might the First World War be called the first industrial war?
2 How did the ability to produce electricity change society?
3 In what ways were the lives of Thomas Edison, Charles Goodyear and Henry Ford similar and different?

⚙ NOTE

1. The acronym ASDIC comes from the Anti-Submarine Division and "supersonics" and was used as a cover name. The Admiralty later made up a false history for the name saying that it came from the "Allied Submarine Detection Investigation Committee."

⚙ FURTHER READING

The rise of the United States as an industrial power changed the course of world history and set a template for the exploitation of inventiveness. One of the foundations of the Industrial Revolution was the textile industry and Giorgio Riello's *The Spinning World: A Global History of Cotton Textiles, 1200–1850* (2009) sets the global context. The first significant war to use industrial-age technology was the American Civil War, when trains, telegraphs, balloons and photography appeared on the battlefield. Barton C. Hacker collects articles on a broad range of topics in *Astride Two Worlds: Technology and the American Civil War* (2016). After the war, the industrialized nations became even more powerful because of the rise of industrial systems such as electricity, as explored in Thomas P. Hughes' *Networks of Power: Electrification in Western Society, 1880–1930* (1993). In particular, Henry Ford became an icon of the new industrialism and has been heavily studied. An easily accessed collection is the PBS American Experience, *Henry Ford*. It includes a digital collection of images, film and articles. Industrial growth was once again thrown onto the battlefield with the outbreak of the First World War. A material culture approach to the war has led to multiple volumes on the objects of the conflict. Peter Doyle in *World War I in 100 Objects* (2014) and Gary Sheffield in *The First World War in 100 Objects* (2013) offer excellent images and supporting text suitable for any level of study.

9 Domestic Technology: Bringing New Technology to the People

Domestic technology has always had the most direct effect on people since it is the technology of everyday life. The Industrial Revolution changed industry but it also changed the relationship of individuals to technology. Prior to the Industrial Revolution, most of the material needs of the household were made by the family or the local community. With industrialization, domestic technology was transformed. The way houses were built, how food was stored and prepared and where people lived underwent major changes. Goods from around the world were brought into the home, and the home, rather than being a place to make things, became the focus of the consumption of goods. In the industrial world, domestic technology contributed to changing ideas about social roles and spaces with suburbs growing up around cities and malls replacing markets.

It is easy to think of technology in terms of big buildings, wars and the mass production made possible by factories. It is certainly true that these things do represent important aspects of the grand sweep of history, but the effect of technology on a more individual and human scale is often a better measure of the changes that technology has made and how technology changes social relations.

Domestic life can encompass a wide range of work. It may include food preparation, care and education of children, making clothes and footwear, manufacturing household utensils, decorative arts, tending the sick and elderly, and the observation of various rituals. How and by whom these activities are carried out has been the subject of a great deal of study by historians, anthropologists and sociologists, looking at our ancestors and present-day social groups. What these studies have revealed is that all cultures have a domestic life, but the roles, particularly as assigned along gender lines, differ widely. Women, by biological necessity, have to be involved in child rearing, but outside of that, societies range from almost complete segregation of the sexes in the domestic sphere to almost complete integration where domestic work was shared. Although it is not the case that all domestic work is women's work, for most of history women's work has been more domestic than men's work. The domestic roles of people are also affected by social standing and economic level, so that the daily activities of the upper class have been different from those of the working class. It is also the case that the status of women in society tends to be strongly affected by the economics of the era. When the economy is bad, women's autonomy is reduced as measured by such things as the right to own property, initiate a divorce and take on paid work, and the degree of domestic responsibility tends to go up. When times are good, women's autonomy may increase, with more property rights, greater economic freedom and lower levels of social control of behavior.

In addition to the distribution of work that people do, there has been a varying range of activities that are considered public or private. As societies became more urbanized and technologically complex, there was a greater separation of the public and private spheres. In forager societies there was almost no concept of a private sphere. In the European Middle Ages, even the wealthy lived and worked in a single great room with family and servants, mixing what we today would consider public and private activities. In some cultures, the home was strictly limited to family members and only men were allowed to go out unescorted, creating a strong division between public and private.

In urban areas, the number of domestic products that were produced in the home declined as civilizations experienced greater specialization of work. The grinding of grain was taken over by mills, footwear was made by cobblers, cloth production by weavers, while clothing was sewn by seamstresses and tailors. As the production aspects of domestic life declined, there was a rise in emphasis on managing household activities, cleaning, and creating a private domestic "haven" away from the public sphere. Notions of proper behavior were based in part on ideas about public and domestic space and what could or could not be done in each space.

❧ The Green Revolution

Food and the home are intimately linked. For most of human history, food preparation was almost entirely a domestic activity, whether undertaken by an individual family or collectively. Many of the most significant changes in global society were based on new processes to gather food. The use of fire transformed food and the beginning of settlements was related to the beginning of agriculture. Each time food production increased there was a corresponding social change as more people were freed from farm labor and made available for other kinds of activities. The Islamic agricultural revolution led to the widespread transfer of plants and animals and, more important, techniques for intensive farming. The changes in agriculture during the proto-industrial revolution era saw a significant rise in agricultural production, particularly in Britain, that in turn helped set the stage for the Industrial Revolution. The industrialization of agriculture introduced the use of machines like tractors, artificial fertilizers, pesticides and irrigation equipment to the farm. These tools and techniques led to large surpluses far beyond local needs and made the movement of people and goods easier. Food security and availability helped to propel the industrial nations to global power.

Successful as industrial agriculture was in Europe and North America, it was not adopted in other parts of the world. In part, this was because places like Mexico, sub-Saharan Africa and India lacked the infrastructure to support industrial agriculture, but it was also because agricultural technology suited to one region could not simply be imposed on another region.

One of the most successful collaborations of science, technology and local interests was the Green Revolution that started in the 1950s and brought new crops and farming techniques to a number of regions, particularly Mexico and India. The start of the

Green Revolution was in Mexico where a consortium of interests came together to try and improve agriculture. The governments of the United States and Mexico worked with the United Nations Food and Agriculture Organization (FAO) and the Rockefeller Foundation. The Rockefeller Foundation had started working with Mexican agronomists in 1943 when they helped set up the Centro Internacional de Mejoramiento de Maíz y Trigo (International Maize and Wheat Improvement Center). The objective was to develop new strains of maize and wheat (and later other crops) that were suitable for Mexico's growing conditions and had higher yields than the older varieties. New wheat varieties were especially successful, and by 1968 more than 90 percent of wheat grown in Mexico had been developed by Mexican scientists. The new agricultural methods also included the use of fertilizers and pesticides and a vast increase in mechanical systems such as tractors and grain elevators.

In 1961, at about the time that Mexico was beginning to get results from the new agricultural system, India was on the verge of mass starvation. Norman Borlaug (1914–2009), an American plant geneticist who was working in Mexico, was invited by the Indian government to help their agriculture. Borlaug, with funding from the Ford Foundation, arranged for a trial of wheat developed in Mexico in the Punjab region. This proved successful and led to a wider use of the new techniques and varieties.

One of the most important additions to Asian agriculture was "Miracle Rice." Creating high-yield rice was one of the main projects of the International Rice Research Institute (IRRI), a joint organization of the Ford Foundation, the Rockefeller Foundation and the government of the Philippines. Surajit Kumar De Datta, an Indian agronomist working in at the IRRI, identified IR8 as a superior variety. Even without industrial farming methods, it produced five times as much rice as traditional varieties, and with fertilizers this rose to ten times. IR8 was adopted in India and throughout Asia.

The Green Revolution, although partly funded by American organizations, relied on local scientists and farmers for its success. It was not without problems, however. The projects associated with the Green Revolution had Cold War political overtones because they encouraged the use of Western technology. Things like tractors and chemicals were given free or sold at low prices by Western governments with the dual objective of helping and spreading influence. The United States and other Western partners had a large advantage over the Soviet Union in agriculture in this period because Trofim Lysenko had virtually wiped out the study of plant genetics in the USSR.

There has also been criticism about loss of biodiversity, environmental damage from chemicals and the carbon footprint of industrial agriculture that relies on tractors in the field and global transportation systems to move the food. In reply to these criticisms, Borlaug said: "If they [environmental critics] lived just one month amid the misery of the developing world, as I have for fifty years, they'd be crying out for tractors and fertilizers and irrigation canals and be outraged that fashionable elitists back home were trying to deny them these things"(Easterbrook 1997: 80).

Another major issue, which continues to this day, is that industrial agriculture is expensive. This has led to the concentration of farming in the hands of large land-owners and commercial interests, as family farms cannot afford to buy the equipment and chemicals necessary for intensive agriculture. The benefit of the Green Revolution is that the planet can feed many more people today than even fifty years ago, but the technology locks in farmers to a particular system that is in turn dependent on other industries, including chemicals and petroleum.

Fireplaces and Cooking

A good starting point for domestic technology is to look at food preparation. Cooking has always been one of the most important focuses for human life. The cook fire was the center of family and community activity and therefore both a public and a private space. Food and socializing have always been closely related, and the concept of hospi-tality and the duties of host and guest are intimately linked to food. For most of human history, homes were little more than a single room with all domestic activities taking place in the same space, and since the preparation of food was a daily and ongoing activity, the rhythm of the day was the rhythm of cooking. Even with the emergence of urban life, the home continued to be built around the cooking fire. Only the very wealthy could afford to build separate rooms for different activities. The modern sepa-ration of food preparation from other spaces in the home did not change the fact that the kitchen continued to be the focus of domestic life. It follows that the technology of food preparation can tell us a great deal about our relationship to technology and to each other.

Changes in the means of cooking in the home can give us a view of the changing nature of domestic technology. For most of human history, the cook fire was simply an open fire, located outside the house in warm climates or during the summer season, and inside in more temperate places and during the winter. When settlements were few and fuel was plentiful, open fires were easy to maintain, but as settlements grew, fuel became scarcer and cost more either in time spent or money. Several strategies were used to deal with problems of fuel and space in urban settings. One of the earliest was the introduction of stone- or brick-lined fireplaces. These appeared in a number of different places in antiquity, ranging from ancient Palestine to China. Fireplaces used less fuel, concentrated the heat and, because the stone and brick absorbed heat, stayed warm for hours after the fire went out or was banked for the night.

Using open flames to cook food was a bit precarious. To create a more consistent heat for cooking, ovens were created. These were originally simply pits that had hot coals placed in them and the food, wrapped in leaves, placed on top and then covered with earth or sand. This cooking method was used around the world and continues to be popular in places such as Hawaii and Thailand. The pit oven and the kiln for making ceramics were essentially the same device and were linked to domestic activity for generations. Around 3500 BCE clay and brick ovens began to appear in the Indus River region, northern China and Egypt. Small and portable ovens made of ceramic date to around 1600 BCE in the Mediterranean basin, and at about the same time ovens specifically to bake bread started to appear. Bread ovens were used a bit differently than other ovens in that they were a chamber (clay, brick or stone) that had a fire lit inside; when the oven was hot enough, the fire and ash were removed. Baking was by the stored heat. The simple design of such ovens meant they could be built anywhere out of local materials.

As stone and brick homes became more common, fireplaces replaced open fire pits for cooking and heat. Early fireplaces were often just a stone or brick structure with a channel or pipe above it for the smoke. These were easier to use than an open fire, but they were not very efficient and generated a great deal of smoke inside the house. Better chimneys helped, but it was not until the eighteenth century that major redesigning of the stove took place. A number of people created stoves or fireplaces specifically for heating, most notably Benjamin Franklin (1706–90), who in 1741 incorporated baffles (channels for air) that moved cool air around the firebox to capture and circulate heat that would otherwise be lost up the chimney. In the 1790s, Count Rumford (the Anglo-American inventor, scientist and occasional spy Benjamin Thompson, 1753–1814) introduced the Rumford fireplace that used a shallow box with angled sides and a much more aerodynamic flue to increase radiant heat and reduce smoke inside the room (Figure 9.1). Both Franklin and Rumford designed their stoves with economy in mind; each design got more heat from less wood.

The introduction of the fireplace and the chimney transformed house construction. Homes were built around the fireplace as the place of food preparation, domestic chores such as sewing, social interaction, heating, cleaning and entertainment. Long after open fires ceased to be the main source of food preparation and heating, many homes still have open fireplaces for their social and decorative function.

Another solution to the problem of fuel was the introduction of charcoal. Charcoal was known in antiquity, and had been made and used by smiths going back to the ancient Egyptians. In parts of the world such as urban China and India where wood was very expensive, cooking over charcoal braziers became popular, and to facilitate fast cooking with the smallest amount of fuel, culinary styles that used small pieces of meat and vegetables such as stir frying and satays became popular. In places where wood was plentiful and inexpensive, the longer cooking times of roasting for big cuts of meat were more affordable.

Cooking over a fire, even in a fireplace, was a challenge. Food needed constant attention, keeping the fire at the right level was difficult, and moving heavy pots and other

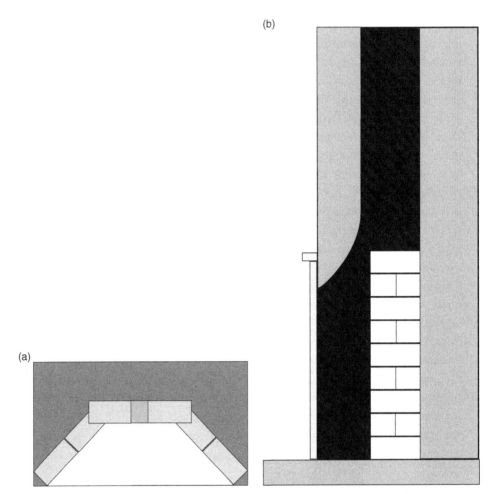

Figure 9.1 Rumford fireplace. The shallow hearth and narrow flue increased heat production and made chimneys much more efficient. They could be retrofitted into existing fireplaces, as shown in the side view (b).

cooking tools required strength. In rich households, fireplaces had hooks, swiveling supports and pot stands. Some even had hoists and crank-operated rotating spits. Even with the various aids, it was hot, tiring work. Although various kinds of stoves had been invented over time, it was really the mass production of cast iron that changed the role of the fireplace. Cast iron stoves became increasingly popular in the late eighteenth and early nineteenth centuries. By 1800 a growing number of houses in Europe and America had cast iron stoves, although many also had a traditional fireplace as well. The earlier industrial-era stoves were not really built with much consideration for the cooks. They were often very low to the ground, had very small flat upper surfaces for pots, and if a compartment for roasting or baking was part of the design, it was often so small as to be of little use.

Since the fireplace had fulfilled a number of roles, including home heating and cooking as well as some other domestic chores such as heating bath water and warming irons, the cast iron stoves were also expected to do those jobs, and this also affected the design. Manufacturers made their stoves increasingly decorative, partly because they were a major item of furniture in the most used room of the house, and partly because they just could, showing off the details they could cast. Some women complained that the decorations simply made the stove harder to clean. Even though it was much more efficient than an open fireplace, there were complaints that the social aspects of the fireplace were being lost, and that the new stoves might contribute to a drop in morals. The writer and social commentator Nathaniel Hawthorne said that cast iron stoves were "cheerless and ungenial," preferring the open hearth for gathering the family (Hawthorne 1970).

In the United States, 102 patents were issued for cast iron stoves between 1835 and 1839. One of the most popular was the "Saddlebags" model produced by William T. James of Troy, New York. It had a large front door, but curved up from the firebox to a wider top, making it look like saddlebags on a horse. Instruction guides for domestic work, such as Mrs. William Parkes (Frances), *Domestic Duties; or, Instructions to Young Married Ladies on the Management of their Households* (1825, 1828 and 1846) with its instructions on using the new stove technology, give evidence that cast iron stoves were increasingly preferred over the open hearth.

It was not unusual in North America to move the cook stove according to the seasons, taking it outside or to a summer kitchen in warm weather and then moving it back inside to heat the house during the colder part of the year. As stoves got larger, this became much more difficult, until those who could afford it began to separate the job of home heating from cooking.

By the middle of the nineteenth century, coal began to replace wood, particularly in larger stoves, but there were drawbacks to coal stoves. The coal had to be low in sulfur or the smoke would affect the taste of food and could fill the house with noxious fumes. Flues had to be better constructed and maintained, but the advantage of coal was the higher temperatures and greater concentration of heat energy. As a heat source, coal became a common furnace fuel used well into the twentieth century.

The first gas stoves appeared in the early nineteenth century and were based on coal gas. James Sharp, who was the assistant manager of the Northampton Gas Company in England, began to experiment with gas stoves, and in 1826 patented a gas stove; by 1834 he was producing gas stoves commercially. A number of gas stoves were shown at the Great Exhibition of 1851, but the biggest restraint was the limited gas delivery systems. It was not until the 1880s, when gas delivery networks in major American and European cities had expanded sufficiently to reach many urban homes, that gas stoves became popular. In many cases, they were leased from the gas companies and were part of efforts by those companies to create a greater demand for gas.

Cooking with Electricity

Electric stoves were introduced in the 1880s, but like the gas stove, they had limited use until the electrification of the major urban centers made it practical to use electrical equipment of all kinds in the home. Electrification was not widespread in either North America or Europe until after the turn of the century, and the network radiated out from the major urban areas. Thomas Edison operated a trial street lighting and commercial electrical system in New York in 1881, and in that same year another Edison trial with public electrification started in Godalming, UK. These trials met with general acceptance and even acclaim, but technical problems as well as the cost of equipment limited electrification until the 1920s, when the cost of production began to drop and demand rose. The problem for domestic consumption of electricity was the paradox of needing to have a complex system in place to get people to adopt electricity, while also needing a certain number of people to adopt electricity to make it economical to build the complex system. This boot-strapping problem was in part solved by the electrical companies targeting municipal governments that were already involved in providing things like gas lighting in public places, roads, and water and sewers systems. Once the power grid was established for municipal use, it was much easier to link in commercial and domestic users.

Microwave Ovens: Serendipity and the Expectation of Inventors

With the establishment in urban centers of the electrical power system, the domestic use of electricity grew steadily. The original electrical device in most homes was lighting, but the electrical companies, hoping to increase the use of electricity, began to introduce more and more electrical devices for the home. Water heaters, washing machines, vacuum cleaners, toasters and radios were all made possible by the electrification of the home. One of the most recent additions to the heating tools of the kitchen is the microwave oven. It was a serendipitous development, with its roots in the electronic tube industry. In 1920, Albert W. Hull (1880–1966) made a tube called the magnetron that produced microwaves. This device was at the heart of radar systems used during the Second World War, and thus of high interest for research. In 1946, Percy Spencer (1894–1970), an engineer working at the Raytheon Company, noticed that a candy bar in his pocket had melted when he was near a working magnetron tube. He tried heating popcorn, and then an egg. He made a metal box to contain the microwave energy and found that it heated food very rapidly. This led Spencer and associate Roly Hanson to commence the secret project they called "the Speedy Weenie." In 1947, Raytheon introduced the first commercial microwave oven, called the "Radarange," designed for large restaurants and institutional use. It was the size of a refrigerator and required plumbing for cooling. Sales were not huge, even after a smaller, home model was introduced around 1952. In 1965, Raytheon purchased Amana Refrigeration and thus acquired a company that understood the domestic market. In 1967, a counter-top model was introduced, and sales took off. Today more than 90 percent of American homes have a microwave oven.

The microwave oven is an interesting piece of technology because it demonstrates the problem of expected use versus actual use of tools. The original domestic microwave ovens were large, often big enough for a whole chicken since they were designed to replace conventional ovens. It seemed logical to the inventors of the original ovens – the microwave was faster and cleaner, and used less energy than conventional gas or electric ovens – that the microwave would be a better choice than a conventional stove. It seemed the perfect choice for the modern, post-war home. The problem was that microwave ovens could not brown, bake, make things crisp or caramelize food, all processes that were possible with conventional forms of cooking. A fully cooked chicken breast from a microwave was essentially steamed and looked like a flaccid white lump, while baked goods required serious food chemistry to produce something that even vaguely resembled a golden-brown cake or a flaky pie crust. Rather than replace gas or electric ovens, the microwave became a tool used primarily to reheat things, including packaged, pre-cooked meals. Surveys suggest that the two most common things microwaved are cold coffee and popcorn. Professional chefs often despise microwave ovens and it is rare to find cookbooks that suggest using them, except to thaw or melt ingredients to be used in more conventional cooking methods. The story of the microwave oven shows us that efficiency and economy are not always the determining factors in the adoption of new devices.

Refrigeration

While the heating of food was a prehistoric part of the domestic sphere, the mechanical cooling of food was a far more recent addition to the home. All cultures that lived in regions with winters used cold to preserve food, but this was a strictly seasonal activity. Wealthy Romans had snow and ice brought down from the mountains during the summer and there were commercial sales of flavored snow. By around 1200 CE, in temperate regions such as northern Europe, ice was collected in winter and stored in ice-houses, often packed in sawdust as insulation, for use during the summer. Although the use of harvested natural ice was common in many places, the commercial exploitation of natural ice and delivery of ice for commercial and home use in iceboxes began at the beginning of the nineteenth century and reached a peak at the beginning of the twentieth century. Frederic Tudor (1783–1864) made a fortune shipping ice from the New England states (Connecticut, Maine, Massachusetts, New Hampshire, Rhode Island

and Vermont) to England, the Caribbean, India and even Australia, using specially designed ships and ice warehouses. The iceman became a familiar figure in the urban landscape and ice deliveries continued as a commercial practice well into the twentieth century. The use of natural ice was largely replaced by manufactured ice by 1920, as artificial refrigeration became available. Refrigeration technology was a product of the Industrial Revolution and the utility of scientific knowledge about heat. In 1805, Oliver Evans (1755–1819) designed the first machine that used vaporization for cooling, while a machine capable of making ice was demonstrated by John Gorrie (1803–55), although he failed to make a commercial success of his system (Figure 9.2).

The origins of the modern home refrigerator really come from the work of Baltzar von Platen (1898–1984) and Carl Munters (1897–1989) on the gas absorption refrigerator. The refrigerator they invented had no moving parts and, in what seems at first glance to be a paradox, used propane, electricity or kerosene for cooling. The gas absorption refrigerator uses a cycle of evaporation (the heating point) and condensation to remove heat from inside the refrigerator and radiate it to the outside air. In 1923 AB Arctic began production, and in 1925 the company was taken over by the industrial giant Electrolux, which expanded the production of domestic refrigerators.

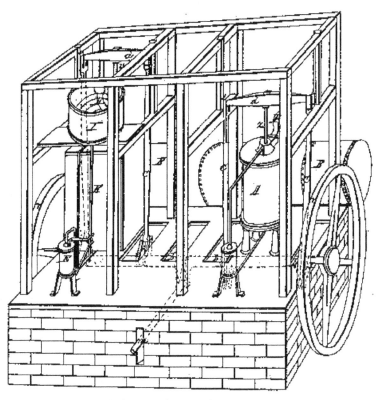

Figure 9.2 Diagram of John Gorrie's ice machine.

In competition with the gas absorption refrigerator were models that used electrically driven compressors. The leader in developing this system was General Electric. Their Guardian units required a separate room for the equipment that was then connected to the refrigerator in the kitchen. It was a luxury item, but part of a larger plan by General Electric to increase the domestic consumption of electricity. The much more convenient Monitor Top refrigerator was introduced by General Electric in 1927 (Figure 9.3). It consisted of a small chest for the food, with the compressor and heat exchange coils on top, covered by a decorative cowling. More than a million of these units were sold, making it a commercial success, although it was still a luxury item. In the original models the gas in the compressor system was either sulfur dioxide or methyl formate, both of which are toxic. In part because of the danger of the coolant, the less toxic chemical freon was introduced in the 1930s by the Frigidaire company. Freon was found to be having a significant effect on the ozone layer of the atmosphere and was banned in most places in 1987.

The refrigerator as a common domestic appliance did not really become widespread until after the Second World War. Following the war, the mass production of refrigerators lowered the price, and the introduction of technical advances such as freezer sections and frost-free systems made them much more attractive. The introduction of

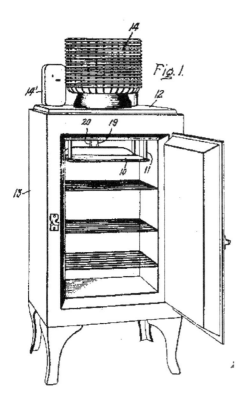

Figure 9.3 General Electric Monitor Top refrigerator patent.

home refrigeration improved food safety and quality, but it also offered a new commercial opportunity since frozen products such as ice, ice cream and other products could be sold into the domestic market. One of the pioneers of frozen food was Clarence Frank Birdseye (1886–1956). Birdseye learned about fast freezing of fish from the Inuit of Labrador, Canada. Things that were rapidly frozen did not produce large ice crystals that would lead to cell damage and poor-quality food when thawed. He went on to invent a way to fast freeze meat, fruit and vegetables. In 1929, Birdseye sold his company and patents to Goldman Sachs and the Postum Company, which would become General Foods Corporation, one of the largest processed food companies in the world. Frozen food offered people the chance to eat things that were not in season, and improved the taste and quality of food over most other forms of preservation such as canning and drying.

✿ The Sewing Machine

Another device in the home that was eventually electrified was the sewing machine. The original idea for a machine that could produce a mechanical stitch to hold materials together is generally credited to Thomas Saint (of whom little is known), who was a London cabinet maker. He was issued a patent for a sewing machine for leather and canvas in 1790, but no actual machine seems to have been produced. Solving the various problems of machine stitching really came from three inventors: Walter Hunt (1796–1860), Barthélemy Thimonnier (1793–1857) and Elias Howe, Jr. (1819–67). Around 1833, Hunt introduced the eye-hole needle to carry the upper thread and a shuttle to carry a lower thread. When the upper thread formed a loop, the shuttle passed through it, creating a locking stitch. Thimonnier used a hooked needle to create his locking stitch, but, unlike Saint and Hunt, actually developed his device to a commercial level when in 1840 he ran a shop in Paris that used eighty sewing machines. Thimonnier's business was attacked by tailors and interrupted by the 1848 Revolution. He moved to England, but he could not make money as his machine had been surpassed by other inventors. Howe was in many ways the most successful of the three early innovators, although his work depended on the same system of eye-hole needle and shuttle as Hunt's machine. Howe received a patent for his device in 1846, but failed to earn much from his work until he successfully sued other sewing machine manufacturers and eventually earned more than $2 million from licensing fees.

One of the people that Howe successfully sued was Isaac Merritt Singer (1811–75). Trained as an engineer, when he saw a sewing machine being repaired, he set about improving the design. In 1851 he patented what we would recognize today as a sewing machine with a reciprocating straight upper needle mounted on a rigid arm. He also powered his machine with a foot-operated treadle. Although Singer had to pay licensing money to Howe, he was enormously successful. His machine worked, was easy to maintain and was rugged – just the thing for families on the move in the great era of settlement in America. The Singer Company continued to introduce new models, including the first electrical sewing machine in 1880. By 1900, Singer sewing machines represented about 80 percent of world-wide sales.

Singer's success was not just mechanical, however. He was also a marketing pioneer, introducing hire-purchase to the domestic market. This system allowed families to buy a sewing machine through a series of payments over time, making the expensive machine available to people of even modest income. Singer also set up sales territories, dividing the United States into regions, and in effect created the franchise business. When Singer died, he left an estate worth $14 million (about $290 million today).

The invention of home sewing also led to the complementary invention of the paper sewing pattern. First created by Ellen Louise Curtis Demorest (1825–98), they were made a commercial success by the husband and wife team of Ebenezer Butterick (1826–1903) and Ellen Augusta Pollard Butterick (1831–71). Patterns from the Butterick Company were printed on tissue paper that could be pinned to fabric to facilitate accurate cutting of the various component pieces of a garment. Using the patterns allowed anyone with a sewing machine to create fashionable and tailored clothing in a huge range of styles. Home fashions were dominated by local customs until the end of the nineteenth century, but two developments helped to transform fashion ideas. The first was the introduction of the fashion magazine, starting with *Harper's Bazaar* (USA 1867), *Cosmopolitan* (USA 1886), *Vogue* (USA 1892) and *Fujin Gaho* (Japan 1905). While these magazines were promoting the purchase of commercially made fashions and aimed at an urban market, they quickly became a source of inspiration for home sewing. The appearance of the motion picture around 1907 was also a source of fashion ideas. The pattern makers such as Butterick (and later such sources as Vogue Patterns and McCall) duplicated the fashions from the news, fashion magazines and films, spreading both the fashions and the conception of a fashion industry.

For the very poor, hand sewing would remain the primary method of making clothes until the twentieth century, but the sewing machine allowed a far greater number of people to make clothes when they were too poor or too far away from commercial sources to buy new garments. The machines also allowed women to make money by sewing, altering and repairing clothing as a form of cottage industry.

Although the sewing machine helped to transform home life for millions of people, the industrial application of the sewing machine was in many ways the culmination of the textile revolution in the Industrial Revolution. The mass production of apparel became possible and the garment industry grew quickly at the end of the nineteenth century. Unlike most of the industries associated with the Industrial Revolution, the garment trade employed large numbers of women, mostly as low-paid seamstresses paid by the piece. Working conditions were often terrible, but the garment industry offered employment to people with few skills or little education. In the United States, the early garment industry was one of the biggest employers of immigrants. The industry would reduce the unit cost of clothing to the point that only the poorest of the poor could not afford manufactured garments. The garment industry today has a global value of over $1 trillion dollars (2017). The dark side of fashion continues to be the sweat shops that remain low-paying and dangerous. Two cases highlight the ongoing problems. In 1911, the factory of the Triangle Waist Company located on the eighth, ninth and tenth floors of the Asch Building in New York City caught fire; 146 people died. It was discovered that the owners had locked the doors to prevent workers from taking breaks or stealing clothing. In 2013, the Rana Plaza building in the Dhaka District of Bangladesh collapsed, killing 1,134 people, including many garment workers. The building had been evacuated when large cracks appeared, but the garment workers were ordered back to work the next day under threat of lost wages or being fired. Although the owners of the building were found to have used poor materials and added three illegal stories to the building, the owners of the textile factories were also blamed for ignoring the safety of workers.

❈ Architecture: A Machine for Living In

While the devices inside the home have transformed the domestic sphere and domestic labor, the very structure of the house has also been transformed, particularly since the Industrial Revolution. As even the working class gained economic power, the home increasingly segregated activities into separate rooms. Kitchens were dedicated to food production, bath rooms were for personal hygiene, bedrooms for sleeping, living rooms for recreation and family gatherings and so on. More rooms meant more space was needed, but in urban areas, particularly in Europe and Asia, such space was either unavailable or extremely expensive. The initial solution was for cities to spread out horizontally, and in North America, where land was comparatively cheap, there was room to grow; but as cities grew bigger, transportation became a problem. People were too far from work to walk and only the wealthiest could afford carriages, so the solution was to introduce mass transport. Cities such as Paris and London created subway systems, and many cities had trams and passenger rail systems.

Another solution was the creation of satellite communities. The development of rail systems made it economical for people to live in towns outside large cities and commute to work. In addition to existing towns becoming dormitory communities, the rail connections led to the building of new suburbs. The garden city movement, pioneered in Britain by social reformer Ebenezer Howard (1850–1928), proposed to build small communities surrounded by green space that were connected to larger urban centers. The original plan was to create a new city based on urban planning that separated industrial and residential space. Howard helped create two cities based on his ideas: Letchworth Garden City and Welwyn Garden City, both north of London.

✿ Skyscrapers

The other solution to the problem of high land cost in urban centers was to build up. With a great demand for land, the value of property in large cities like New York, Chicago (the fastest growing city in the world at the beginning of the twentieth century), Paris and London encouraged the greatest density possible. That really meant building up, but two factors placed practical limits on building height. The first was obvious to anyone who has climbed to the top of any tall structure; climbing stairs was tiring and moving supplies, furniture and other materials was a challenge. The second was less obvious, and that was the problem of getting water to the upper floors of the building. Any building that was to be used by people for housing or offices needed water for bathrooms, drinking and firefighting.

The problem of water was solved by the introduction of high-pressure pumps such as John G. Appold's (1800–65) impeller pump invented in 1851. These pumps were made more practical when electrical motors became commercially available, as well as architects designing buildings with water tanks at the top of the building that then used gravity to supply the plumbing system.

The problem of moving people and materials was solved by the introduction of the elevator. Various forms of lifting platforms had existed back to the time of the Romans, but they tended to be dangerous and rather labor intensive, requiring constant attendance. In 1852, Elisha Otis (1811–61) invented the safety elevator (Figure 9.4). This consisted of a rope or cable to move the platform, and a spring system that released steel clamps if the rope was broken. Otis formed the Otis Steam Elevator Works, and his big break came in 1853 when he did public demonstrations of his safety elevator at the New York World's Fair. The first passenger elevator was installed in the E. V. Haughwout

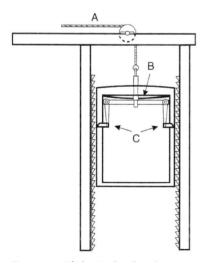

Figure 9.4 Elisha Otis's safety elevator patent drawing from 1861.

Table 9.1 Early skyscrapers.

Date	Name	City	Height (m)	Height (ft)
1889	Auditorium Building	Chicago	82	269
1890	New York World Building	New York	94	309
1894	Manhattan Life Insurance Building	New York	106	348
1899	Park Row Building	New York	119	391
1901	Philadelphia City Hall	Philadelphia	155.8	511
1908	Singer Building	New York	187	612
1909	Met Life Tower	New York	213	700
1913	Woolworth Building	New York	241	792
1930	Chrysler Building	New York	282	925
1931	Empire State Building	New York	381	1,250

Building in New York City in 1857. The great German inventor and industrialist Werner von Siemens (1816–92) created the first electric elevator in 1880.

The first skyscraper, the Home Insurance Building, was completed in 1885. The building was ten stories tall, which was not monumental in terms of overall height, but it was a landmark in architecture for the use of an interior framing system rather than the building's weight being carried by the exterior walls. The framing was a series of steel and cast iron columns and beams. This was the only practical solution to the problem of the thickness of the base of large structures, since in conventional design the floors had to span the space from exterior wall to wall. Thus, the taller the building, the wider the load-bearing walls had to be, often reaching several meters thick. The use of interior framing was not new to the Home Insurance Building, since various forms had been used in England, including the Crystal Palace in 1851, but it became the standard for large-scale construction. A kind of race to the top followed (Table 9.1).[1]

✿ Cities and Cars

With the introduction of the low-priced automobile, more people could live away from work and services, and this in turn created the first car-based suburbs. In large part, these suburbs were relatively close to the main urban area and were filled with relatively wealthy families who could afford cars. In many cities, the first quarter of the twentieth century saw a slow exodus of middle-class families from the inner cities and away from the industrial zones. This is not to say that there had not been rich and poor areas of cities, but the coming of the car meant that a much greater physical distance could separate people.

The great explosion of the suburbs occurred after the Second World War. In Asia and Europe, suburban growth was delayed by the devastation of the war, but in the United States, and to a lesser extent in Canada and Australia, the movement of people to cities was almost immediate. Unlike the demobilization following the First World War, the returning soldiers did not go back to the farms but gravitated to the urban areas for work. In the United States, the Servicemen's Readjustment Act of 1944, also known as the G.I. Bill, offered money for education, and low interest loans for mortgages and to start businesses. The soldiers, often trained during the war to do everything from welding and truck driving to flying huge aircraft, found employment, bought cars and moved to the suburbs in huge numbers.

Houses built for the middle class before the Second World War tended to replicate a simplified version of a manorial house. The main building was on the street with a porch that provided a transition from public to private space, while the garage was at the back of the lot, as far from the living quarters as possible, just as stables had been located in an earlier age. Access to the garage was through an alley at the back of the home that also served as access for garbage collection and deliveries of materials like coal, and as the entrance for workers. This design was expensive in terms of land use, and as the automobile became more important to urban and suburban life, the placement of the garage shifted to the side of the main house as an attached garage or car port. Less land was needed, since there was no alley to build and service, and the car was accessible without going outside. The side garage meant that lots were wide, and this meant developers could not place as many houses on a block as they wanted. The ultimate suburban lot plan placed the garage at the front of the house, close to the street. Rather than an approachable front entrance, the main street feature of the modern suburban house was a giant garage door.

There are suburbs around every city in the world, but the problems remain the same. How does the local population balance the value of accommodation against the cost in time and money of transportation to and from the city? Particularly in Asia and Europe where land is expensive or its use restricted, the form of suburbs differs from those in the Americas and Australia. In Japan, rail transport is widespread and people walk to the stations from small houses or apartments. Despite being a major producer of automobiles, in Japan cars are expensive to own and there is little space provided for them, even in the suburbs. Housing cost and availability push many people out of the urban areas, but good commuter systems mean that they can continue to work in the urban centers.

Table 9.2 Population density for Beijing and its suburbs.

District	Population density per km²
Beijing	24,685
Inner suburbs (4)	7,259
Outer suburbs (6)	915

Megacities (cities with populations over 10 million) are often accompanied by mega-suburbs. For example, Beijing has four "inner" suburbs and six "outer" suburbs with a population over 15 million, around the central city with a population of around 2 million. The suburbs are connected by roads and the Beijing Subway system, but the city as a whole is so large that there is more employment in the suburbs than in the central city. Most people in Beijing and its suburbs live in apartments and take transit to work. Traffic congestion, once rare in Beijing because automobiles were rare, is now a major problem as the increasingly affluent population see cars as both transport and an indication of social status. The big distinction between the city and the inner and outer suburbs is population density (Table 9.2).

"The house is a machine for living in," said the architect Charles-Édouard Jeanneret-Gris, better known as Le Corbusier (1887–1965) (Le Corbusier 1923). Le Corbusier was one of the pioneers of the Modernist movement (sometimes known as the International style) in architecture. Modernism tended to celebrate the beauty of clean lines, the use of cutting-edge materials and building techniques, and the power of the machine. The Modernists rejected the decorative embellishments that had been found in most forms of architecture prior to the end of the nineteenth century, and tried through a reconsideration of the use of space to bring rationalism and scientific ideas to buildings. They used structural steel, concrete and new materials like plastics and plywood. Although Modernist homes continue to be a favorite subject for architectural magazines, they were less popular with the general public, who found them a bit cold, austere and often lacking storage space. Yet some of the ideas of the Modernists have transformed skylines as most skyscrapers reflect Modernist esthetics, and the use of highly engineered materials such as structural steel, plywood, plastics and poured concrete.

✿ Mass-Produced Houses

The intersection of industrial methods and domestic life can be seen in the homes of Levittown, a suburb of New York City. The development was the idea of William Levitt (1907–94), a builder who recognized the huge demand for housing following the Second World War. In 1947 Levitt and Sons built 2,000 rental homes at Nassau

County, New York, on Long Island. The houses were all rented in two days, and the company added another 4,000 houses to the development. In 1949, Levitt began to sell what they called "ranch houses," with a planned base price of $7,990 (about $81,740 in 2015), inexpensive even by the prices of the day. Five models were offered, although the differences were largely cosmetic. Included in these mega-developments were schools and parks. By 1951, Levitt and Sons had built more than 17,000 houses in planned communities in New York, Pennsylvania, New Jersey and Puerto Rico.

Levitt kept his costs down because he applied an assembly line approach to building houses. The system used teams that did just one job, moving from one building site to the next, so that framers just framed and they were followed by the plumbing installers, the electricians and the finishers. They used pre-cut lumber from their own factory for framing and common components for each house; windows, bath tubs, doors, sinks and so on were all the same (although they might be different colours), were interchangeable and were purchased in bulk. Each length of pipe, run of electrical wire and square of flooring was calculated and accounted for in the course of construction.

The Suburbs: Between Techno-Dreams and Isolation

Domestic technology and the automobile had combined to create what many people desired – a modern, self-contained home that had more space and amenities that families only a generation or two earlier would have associated with extreme wealth. Even entertainment was self-contained, with radios, televisions, stereo systems and later video players and computers bringing the world into the living room, bedroom or even bathroom.

It also proved to be something of a trap. Suburbs were so large that it wasn't just getting to work that required a car or long transit ride, it was getting the children to school, traveling to appointments, lessons, entertainment and sports. Every item of food had to be collected by transport, and while the variety of tins and packaged and frozen foods might mean that a family only had to shop once or twice a month, it also meant that the sense of community that had grown up from the time of the Greek agora through the neighborhood shops of the old cities was gone. Increasingly, people felt more connection to characters they saw on television than they did with the people next door.

One of the changes to social life that came about because of the Industrial Revolution and the rising power of the middle class was the creation of a domestic ideology based on rigid conceptions of proper spheres of activity for men and women. The segregation of the sexes has, through history, ranged from strict activity and physical separation to almost complete integration. In most cases, the degree and justification for the separation is based on some claim about the ordained or natural order. In the post-war industrial world, the suburbs represented a kind of economic and social separation of the sexes, as women left or were forced out of jobs they had done during the war, and the economy was strong enough that many families could afford to have only a single person working for pay. The "nuclear family," being a family made up of a father, mother and their children but no extended family members like grandparents or aunts, became an increasingly standard model for family life in the industrialized countries by the middle of the twentieth century.[2] Media tended to reinforce this social organization, and the suburbs seemed to be a physical embodiment of the nuclear family.

There was a reaction to the expansion of suburban life. Starting in the 1950s and building in the 1960s, criticism of suburbs portrayed life there as dull, stultifying, racist and the perfect icon of conservative conformism. The suburbs were marketed as the perfection of the middle-class dream of property and domestic contentment, but people often found them a boring and homogeneous world in an unending maze of identical buildings. In North America, the stress of suburban life led to the use of prescription drugs like Miltown (meprobamate) and Valium (diazepam) which were prescribed to millions of women. This was satirized by the Rolling Stones in their 1966 song "Mother's Little Helper." Movies about life in the suburbs such as *Strangers When We Meet* (1960), *Bachelor in Paradise* (1961) and *Bob & Carol & Ted & Alice* (1969) ranged from light comedy to dark tragedy. In 1975, Ira Levin wrote *The Stepford Wives* (later made into two movie versions in 1975 and 2004), that told the story of an idyllic suburban community where the women have all been replaced by "gynoids," robot-like replicas of the original people.

One of the foremost commentators on the structure and meaning of cities and suburbs was Jane Jacobs (1916–2006), whose 1961 book *The Death and Life of Great American Cities* continues to influence urban planning around the world. Jacobs made the argument that modern urban planning was based on the same kind of scientific management that had been applied to factory work, that cities were in fact a kind of machine. To make cities "rational" meant cleaning them up (urban renewal) and separating the functions, so there would be industrial areas, commercial zones and residential enclaves. These areas would be connected by high-traffic roads and highways that allowed easy access to the work zones from the residential zones. Jacobs argued that this "rational" system actually destroyed communities, stifled innovation, isolated individuals, and ultimately led to the economic decline of American cities. Although ideas about "scientific" urban planning affected cities around the world, the cities that are today considered the most attractive tend to mix housing, recreation and commerce rather than segregate them.

Domestic technology, or the systems and devices most closely associated with everyday life, has transformed human organization. The central problems of domestic technology were providing food, shelter and social support. Changes in the ways these three things were acquired were brought about by external conditions such as the introduction of electricity, but domestic concerns were also the driving force in promoting lifestyle in the wider society. The demand for private housing in the industrialized world, for example, was a product both of the rising economic power of the family so that dreams of home ownership became possible, and of the decline in the cost of building and using a private home that came from industrialization, transportation and the networks that supported everything from food distribution to municipal services such as sewers and street lights.

By the middle of the twentieth century, the "home" became a very valuable economic target for innovation and sales. Consider, for example, the number of commercials on television that have something to do with making the home better in some way: cleaner, more secure, better smelling, easier to store and prepare food, and a place to sleep better. Each advertisement claims that a form of technology (often "new and improved") will make life better in some way. A cynical view of this suggests that advertisers often attempt to make people dissatisfied with what they have as a means of getting them to buy something that they don't really need, but this is not as straightforward as it seems. Although there is often a general sense of nostalgia about the past, it is the general case that the materials available today are superior to the materials available in the past. The laundry soap available today really does clean better than the soap available twenty-five or fifty years ago.[3] Home electronics are more reliable. Even something as humble as house paint is today safer, less toxic and easier to apply, and comes in an infinite range of colors.

Whether things work better today or in the past is not the real question about the role of domestic technology. The real question is what the transformation of the domestic sphere by technology has done to personal and community relationships. The family was at the heart of the creation of civilization, and the spaces people occupied were designed to support the structure of the family. As civilization gave us specialized occupations and the economic power to create separate spaces for different activities, such changes took place as much in the home as in the marketplace. Specialized spaces coincided with the creation of public and private spheres and the ideologies that controlled who could be where. As well as personal relations, we can see that a community of people who live in rented apartments in an urban area will have different concerns than a community of people who live in private homes in the suburbs. The suburban shopping mall may seem superficially like a modern version of the ancient Greek agora, being a place where people meet and shop, but it is in fact a private space and not really open for public activity. When public space disappears, the interface between the private and the public, the domestic and the commercial, also disappears.

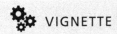 VIGNETTE

Industrialization and Gender Roles

Is technology gendered? In other words, do we consciously or unconsciously embed conceptions of a distinction between the genders in the devices and systems that make up our technological world? While an argument can be made that a hammer or a laptop computer does not know or care whether it is being used by a man, a woman or a Martian, the object is not the technology. The hammer or computer came into existence and is used in a certain way because of a series of decisions about design, finances, marketing, regulations, education and social expectations. When we present technology this way rather than as a series of artifacts, it becomes much clearer that gender is embedded in technology. Women make up only a small part of boards of directors in major corporations, governments and regulatory bodies. Women are only starting to have a presence in engineering, computer science and the automotive industry. Men were, until recently, the test subjects for most medical and biophysical research. It is not surprising to find that objects like hammers and automobiles are designed around the average-sized man or that medical treatment is often based on the expected outcome for men. We can take the ancient Greek philosopher Protagoras' statement "Man is the measure of all things" in a quite literal sense when it comes to the manufactured world.

We can also look at gender as an issue in the participation of men and women in the use of technology in the public sphere. In this we see the place of women vary wildly over time. In the agrarian world, the distinctions about gender and work were largely meaningless: everyone worked and only the wealthy could afford to use gender as a division of labor. This did not mean that there weren't different social expectations of men and women, but the central activity of farming had to be undertaken by everyone. In some periods women were allowed to have a profession such as weaver, potter, brewer or business owner. Then, often associated with periods of economic downturn, women are forced out of positions where they have created and used technology. The coming of the Industrial Revolution was a boon to many women who could gain a degree of freedom by working in the factories. Even if their wages were only a fraction of what men made, it was better than the virtual slavery of subsistence agriculture. By the end of the nineteenth century, women who worked in the factories were largely streamed into specific industrial jobs, such as in the textile industry. Even today, the global textile industry depends on armies of low-paid women to sew our clothes. Women were excluded from heavy industrial work such as steel making, the automotive assembly lines and construction and thus also excluded from higher-paying work.

Various leaders and social commentators throughout history have argued about what were appropriate activities for men and women. Those roles were enforced by social rules that ranged from unwritten general expectations of social norms to laws that defined specifically what men and women could or could not do. Ironically, the introduction of new tools and products removed any justification for having gender differences in access to technology and work that were not based on religious or social doctrine. Better healthcare at childbirth and the 1867 introduction of baby formula by Justus von Liebig (1803–73) reduced the necessity of long periods of nursing. Reliable birth control, starting with latex condoms in the 1920s and the introduction of oral contraceptives in 1960, placed control of reproduction in the hands of women. The introduction of mass public education in the late nineteenth century chipped away at the educational limits on women.

The other great distinction that restrained women from complete integration in the work world was physical strength. This constraint was frequently a gross generalization, since there have always been strong women and weak men, but jobs that required high levels of physical strength such as iron working and firefighting were traditionally limited to men. Yet one of the most important objectives of invention has been to reduce the necessity of muscle power in production. Everything from the invention of the crane in ancient Greece to the introduction of power steering in the automotive industry has reduced the need for muscle power. When necessity opened the doors of heavy industry to women during the Second World War, it was clear that raw muscle power had little to do with modern industrial work and it has even less place in the digital age of production.

Recognizing and addressing gender issues in technology could mean that artifacts such as hammers are made for women, but a more important change would be the inclusion of women and other under-represented groups in the systems that create and maintain our technological world. Participation in the creation and use of technology offers the power of self-determination, freedom from dependence and a sense of value.

The Paradox of Domestic Technology

In the industrialized world, the increasing appearance of machines in the home was promoted as a way of both improving lifestyle and easing workload. In 1983, the scholar Ruth Schwartz Cowan published *More Work for Mother: The Ironies of Household Technology from the Open Hearth to the Microwave* (Cowan 1983). She pointed out that every time a new device was introduced to the household, it promised to save time and improve the quality of life, but in reality women were spending more time on domestic activities rather than less. As cleaning tools such as vacuum cleaners and clothes washers became standard equipment in the home, not only were tasks that had been performed outside the house or at infrequent intervals becoming part of the regular routine of the management of the household, but standards of domestic activity were rising. Rather than beating the rugs once or twice a year, the carpets had to be vacuumed at least once a week. Dirt was not just to be avoided, it was a sign of moral failure and a lack of hygiene. Tidy was not clean enough; with modern domestic tools, the homemaker had to aim for a sterile, germ-free environment. In addition to the heavy equipment, women were sold massive chemical arsenals of cleaners, specialized soaps and detergents, odor controllers, drain cleaners and disinfectants. Ancient alchemists would envy the chemical stores of many modern middle-class households.

✿ The Meaning of Domestic Technology

The greatest transformation of the home in the industrialized world was accomplished by the successful establishment of networks and infrastructure that made the mechanization of the domestic sphere possible. Water and sewage systems were built, electricity and gas networks established, and delivery systems for coal, heating fuel, ice, milk and mail became common. Transport, first by rail and later by automobile, connected people to work and services and contributed to the growth of cities and suburbs. The transportation systems required tracks and roads, power systems and gas stations. Electricians, plumbers, lawn and garden workers, painters, home décor experts, repair people and an army of other support workers appeared to tend to the needs of the modern home. Over time, people were linked into the vast networks that made modern homes possible.

There was, however, a more subtle and deeper effect of the technological household. It transformed the domestic sphere from a place of production and consumption to a place designed around constant consumption. The home was once a place where food was processed, not simply prepared for meals, where clothing was manufactured, soap made, tools built and repaired. Communal activities that often took place outside the home such as washing clothes were reduced and replaced with individual activities contained within the house. Although skills like canning continue to be passed down from generation to generation, they are ancillary activities – the modern family does not depend on them for survival. In other words, the home is now a training ground for the use of technology. It inculcates us in the utility and use of technology and places us within the networks on which the home depends.

1 How did the design of the home reflect the changing forms of domestic technology?
2 What technologies made the suburbs possible?
3 In what ways did ideas about social relationships shape the physical environment of the home?

✿ NOTES

1. Other buildings would rise to great heights, but it would not be until the 1970s that new building materials and techniques would reignite the skyscraper race, starting with the World Trade Center in 1972 at 442 m (1,368 ft). The World Trade Center was destroyed by terrorists in 2001.

2. The term "nuclear family" meaning the minimal functional unit of the family was first used in 1924 in England, but only became common after 1947, giving it a touch of Cold War modernism during the age of nuclear weapons.

3. The secret was in the development of enzymes that could break down stains, another example of the laboratory coming into the home.

✿ FURTHER READING

Histories of technology often favor "big" technology such as railways and weapons. In practical and personal terms, domestic technology touches more people and has a more significant economic impact. An intimate look at cooking comes from Priscilla J. Brewer's *From Fireplace to Cookstove: Technology and the Domestic Ideal in America* (2000). Ruth Schwartz Cowan, a leading scholar on domestic technology, introduces the study in "How We Get Our Daily Bread, or the History of Domestic Technology Revealed" (1998) and her seminal work *More Work for Mother: The Ironies of Household Technology from the Open Hearth to the Microwave* (1983). Where we live and how we live are examined by Richard Harris, and Peter Larkham in *Changing Suburbs: Foundation, Form and Function* (1999). A more popular work, by Margaret Lundrigan Ferrer and Tova Navarra, is *Levittown: The First 50 Years* (1997). Bridging the public and private space, Joanna Merwood-Salisbury's *Chicago 1890: The Skyscraper and the Modern City* (2009) examines the physical transformation of the modern city.

10 The Second Industrial Revolution and Globalization

1763	First public schools, *Volksschule*, in Prussia
1847	Factory Act in Britain introduces ten-hour work day
1927	Transatlantic telephone starts Movie *The Jazz Singer* uses sound technology Charles Lindbergh flies solo across the Atlantic
1929	Beginning of the Great Depression
1933	Adolph Hitler comes to power in Germany
1936–9	Spanish Civil War
1939–45	Second World War Germany uses Blitzkrieg New weapons include radar, jet aircraft and rockets
1945	USA uses nuclear weapons on Japan

The size, range of power and complexity of machines increased dramatically in the twentieth century. With new communications systems such as the telegraph and the telephone, centralized control of business and government reached new levels. The necessity of training to live in the new technological era contributed to the movement for mass education. Education was also needed to create and maintain equipment that was too complicated to operate without specialized knowledge. As a result of industrialization, the economy became global and so did the conflicts. Europe fell into a war that was fought with a new generation of tactics and increasingly powerful weapons. The Second World War was the first conflict to see significant technological advances during the actual course of the war, with codes, aircraft, radar and rockets contributing to the outcome. The ultimate weapon, the nuclear bomb, could only be manufactured by nations willing and able to devote significant industrial resources to the project.

In the era of colonialism, access to global resources had taken Europeans out to the far reaches of the world to bring those resources back to Europe. By the twentieth century, the push for industrialization and changes in transportation created a global marketplace where industrial production and consumption spread from Europe and North America to the wider world. In many ways, the start of the twentieth century began with the end of the First World War. The war had been fought using twentieth-century weapons and nineteenth-century strategies. It demonstrated in the most graphic way possible the power of mass production and the application of modern scientific and engineering concepts to the task of mass destruction. It also demonstrated that industry and politics were global as people and materials were moved around the world to fight the war.

The war also established a new set of relationships among the industrial countries. Britain was battered but still strong. France suffered both territorial damage and economic problems. Even though Germany was not invaded in the course of the war, it lost territory and suffered massive economic and social damage. Germany also experienced a period of occupation by France. Russia, so rich in resources but so poor in industrial infrastructure, was undergoing the Russian Revolution that would eventually create a communist state and isolate the country for more than two generations. In Asia, Japan, who had an alliance with the Entente Powers and aided the British Navy in the Pacific, had seized German territory in the region. Even before the war, Japan had defeated Russia in the Russo-Japanese war of 1904–5. These successes led to the decision to invade China in 1937. Quietly, almost unnoticed because it had been engaged in only limited international trade prior to the First World War, the United States emerged as the strongest industrial power in the world. The North Atlantic region had become the technological and industrial hub of the world.

Major transformations in industry have always led to winners and losers, but new types of jobs and the expansion of commerce have generally created more jobs than were lost. This was not obvious to the people swept up in the periods of rapid change, but the trend in the industrial world has been greater populations and more jobs. The weaver who fell into poverty because of the mills or the ditch digger replaced by the steam shovel might not benefit from the changes in technology, but their descendants lived in a transformed world. The life of a peasant in thirteenth-century Europe was little different from that of the peasant of the sixteenth century, but a person born in 1890 could have traveled by horse-drawn carriage and lived to see a man walk on the Moon.

There were many long-term consequences to industrialization. Larger populations could be supported, but the number of children per family started to decline in every industrial country while the average age of mortality was slowly rising.[1] This was in part due to the general increase in medical knowledge, and the availability of doctors, nurses, new drugs and hospitals, but it was also about a social change that would shape the twentieth century. As governments began to recognize that industrial capacity was important to the national interest, and there was political value in protecting the workforce, the role of government in people's daily lives began to change. Social activism also helped shape the new political and social milieu. Among the social reform movements that started in the nineteenth century were the anti-slavery movement, opposition to child labor, women's suffrage, alcohol prohibition, and medical care groups such as the Red Cross. Some of these were linked with religious organizations, but these movements increasingly transcended particular groups and even crossed national borders.

✿ Labor

Another aspect of the industrial social movements was the organization of labor. Unions, although only indirectly related to the old guild system, nonetheless worked for some of the same goals, particularly to improve working conditions in the very dangerous mines, foundries and factories. The unions also sought to reduce the length of the working day and week, campaigning for the ten-hour day, introduced in Britain as the Factory Act of 1847 that limited the hours of labor of women and children, but which later became a standard for a large range of workers. Starting in the 1860s there were further efforts to limit the hours of labor to eight hours a day or forty hours per week. The Eight-Hour Day movement was part of labor struggles around the industrial world, and was adopted in some places and by some employers (such as Henry Ford, who introduced an eight-hour day in 1914), but it would not be a general rule in the United States until 1938 with the creation of the Fair Labor Standards Act.

Higher pay and less arbitrary treatment were also objectives of the labor movement. Organizations such as the American Federation of Labor and the Trade Union Congress in Britain struggled, sometimes in pitched battles, sometimes by negotiations and popular persuasion, for the rights of workers. The early labor movement was an amalgamation of workers' rights, social reform and political action. The political response to industrialization gave rise to new political groups and the revision of old ones. Among the new groups were the socialists, a term that in the nineteenth and early twentieth centuries covered Marxists, communists, social democrats and even anarchists. In the long run, unions gained power for workers in the workplace, but tended to give up direct political and social action to avoid being labeled as dangerously radical.

Unions are sometimes portrayed as Luddites because they oppose the introduction of new tools of production if those tools eliminate jobs. New technologies tended to create new kinds of jobs and this caused a shift in employment such as the move from agrarian to factory work associated with the Industrial Revolution. Such changes often left a generation of people destitute but following generations better off economically.

Yet there is a technological paradox inherent in the labor–owner relationship in the age of automation. As workers are removed from the production system by the use of automation and robotics, a significant number of people will no longer be able to afford to purchase the products the automated factories produce. New types of jobs such as computer programmer have been created by the shift to the automated and digital economy, but one of the primary objectives of automation is to reduce or eliminate the need for human workers in *all* sectors of the economy. In the twentieth century, the children of displaced factory workers could find work as bus drivers and bank tellers, but those jobs are disappearing as automated teller machines (ATMs) and self-driving vehicles are brought in. Jobs in the next tier such as managers and computer programmers require significant education which may not be affordable or even available in many places around the world. This exacerbates the distinction of the winners and losers in the application of technology.

✿ Mass Education

The most powerful of the invisible technologies is education. Education is a technology that allows us to use and develop other technologies.

For most of history, the majority of education was informal and based on teaching children the necessary skills to work and live in their local environment. Formal education was reserved for the wealthy and a small scholarly class that was generally made up of teachers/priests. Schools as independent places of learning have existed since ancient times with Plato's Academy (founded around 387 BCE) and the Pushpagiri school in India (founded c. 200 CE), but there was only a handful of such organizations. Universities and other places of higher education such as the madrasas began to appear in the tenth and eleventh centuries, but they were elite organizations designed for students who had already mastered the skills of reading, writing and basic mathematics.

The idea that the general public should receive formal education was often proposed, but the first significant effort to create a state-funded elementary education system was in Prussia. In 1763, Frederick the Great decreed that children aged five to twelve should be educated and he established a national education system. The *Volksschule* (literally "peoples' school" but really meaning compulsory elementary education) taught reading, writing, Christian education and music, adding subjects such as mathematics and history later. Prussia offered free elementary education to everyone, built schools and trained teachers. It also established secondary schools and, starting in 1812, all secondary students had to take a final examination, the *Abitur*. High scores opened the door to university education or good jobs in business and the civil service.

The Prussian model was not adopted in other places until after the American Revolution (1765–83) and the French Revolution (1789–99). The new governments wanted to instill a sense of nationalism and foster useful skills, but there were problems. Education was expensive, and in the United States it was controlled by the states, not the federal government. Public education began to be introduced in Japan in the 1870s, following the Prussian model, but it was not until the 1880s that France and other western European states instituted public education, followed by Britain in the 1890s. The United States did not have country-wide public education until 1918.

With the advent of public education, there was also a growth in educational theory. Most elementary education was based on memorization and rote learning, often using state-approved textbooks. This approach was predicated on the idea that children were undeveloped adults. Teaching was enforced by strict and often physical discipline to discourage lack of attention or obedience. The philosopher Jean-Jacques Rousseau had argued that children were not adults in miniature and grew to become adults like a seed grew to be a tree. People like Maria Montessori (1870–1952) and Rudolf Steiner (1861–1925), who founded the Waldorf schools, introduced different teaching philosophies based on the observation of child development. To try and measure development, in 1904 the French Ministry of Education asked Alfred Binet (1857–1911) to devise a method to test children that would identify their level of intellectual development or mental age. In collaboration with Théodore Simon, he produced the Binet-Simon test in 1908. Although the test was supposed to be a scientific measure of intellectual development, it was seen by many as a test of intelligence and therefore the first IQ (intelligence quotient) test.

There was a strong correlation between levels of education and levels of regional success in the twentieth century. Those regions with high levels of education, particularly measured by levels of literacy, had stronger industrial growth, more competent bureaucracies, better health care and better-trained military forces, and were more innovative. The reverse is also true: regions with low levels of education had lower levels of industrialization, poorly run bureaucracies, poorer health, less well-trained military forces and lower levels of innovation. Education, as a technology, seems to come very close to technological determinism, where having public education leads to having a society that can produce and manage complex technology.

✿ The Desire for the Future

As the First World War ended, the future world seemed to be closer than ever, and it was being created by people who believed so strongly in progress that they rejected ties to the past. Henry Ford, as one of the most powerful industrialists of all time, interviewed in 1916 by a reporter from the *Chicago Tribune* said: "History is more or less bunk. It's tradition. We don't want tradition. We want to live in the present and the only history that is worth a tinker's dam is the history we make today" (May 25, 1916).

Ford's view was held by many, and not just in the United States. The Great War of 1914–18 was seen by many people as a fight left over from the age of Napoleon and fought for irrelevant and outdated monarchs who held power only because of the vagaries of history. It achieved nothing but the death of a generation of young men and introduced a kind of mechanized killing. It was history dragging people down, and the only way forward was to embrace the future, and that meant new technologies. One of the characteristics of the new era was the degree to which the locus of invention moved to the United States. Everything from Scotch Tape to FM radio was introduced in America in the period between the wars. Robert Goddard published his "Method of Reaching Extreme Altitudes," in *Nature* 1920, suggesting that a rocket could reach the Moon, and making real the dreams of Jules Verne. In 1927, Charles Lindbergh thrilled the world by flying non-stop from New York to Paris. Inventors were no longer just innovators, but became synonymous with the industrial development that grew up around their inventions. Ford, Edison and Bell were not just inventors, but business leaders and symbols of the new man in the new age.

In the new age of invention, even information was moving faster than ever, with the introduction of commercial radio when Westinghouse-sponsored station KDKA began broadcasting in October 1920. The transatlantic telegraph had been one of the great technological feats of the nineteenth century when the first messages began to flow in 1866, and in 1927 the transatlantic telephone service started, although it used radio rather than cable. Newspapers often had two editions a day, and newsreels from companies such as Pathé, Hearst Metrotone News and Fox Movietone News provided images and sound recordings from around the world. Starting in 1927, the sound for films was produced on the film itself rather than as separate recordings. Frequently cited as the first "talkie," Al Jolson's *The Jazz Singer* (1927) was not the first to use the new technology commercially – that distinction went to a Fox Movietone newsreel of Charles Lindbergh's take-off for his solo transatlantic flight. *The Jazz Singer* was, however, the first box office smash hit using the new technology and generated more than $2.625 million ($366 million today) in earnings. New technology could be very profitable.

People in New Jersey might have gone to see *The Jazz Singer* by traveling through the Holland Tunnel linking Manhattan, New York and Jersey City, New Jersey, under the Hudson River. Started in 1920, the tunnel was named after Chief Engineer Clifford Holland, who died in 1924 before the project was complete. The tunnel was designed specifically for automobile traffic, including a powerful ventilation system to expel the exhaust fumes that would be produced during the 2.5 km (2¾ mile) trip. The automobile was the central feature of the growing American cities, and from the 1920s municipal, state and national planning increasingly favored road construction over other forms of transport. In 1925, the architect and land developer Arthur S. Heineman opened the first motel, the Milestone Mo-Tel in San Luis Obispo, California. The name combined the words "motor" and "hotel," and he charged $1.25 per night.

With the war ended, the soldiers returned to their homes. Many went back to the farms, but the world was not the same. On the surface, there was a huge sense of relief that the war and the influenza epidemic had passed. Heineman's Mo-Tel quickly faced fierce competition as "tin-can" tourists began using their cars to travel across the United States. The 1920s became famous for flappers, parties and, in Prohibition-bound America, speakeasies and jazz. Yet under the surface, the cost of the war in monetary and human terms was immense. Germany fell into a severe economic depression, while France and Belgium, attempting to claim their reparation penalty from Germany, invaded the Ruhr valley, occupying the industrial and resource-rich zone during 1923–4. The influential economist John Maynard Keynes, an important member of the British Treasury at the time, argued that if Germany was allowed to collapse, Britain (and in fact much of western Europe) would be dragged down as well. This set Britain and France at odds, and, combined with the civil unrest in the Ruhr region, the occupation was ended.

❧ Making Records in the Air: Lindbergh and the Promotion of Technology

In 1926, Richard Evelyn Byrd and Floyd Bennet made headlines when they became the first people to fly over the North Pole. Although it is likely that Byrd and Bennet only made it part of the way (their Fokker F-VII could not have traveled the whole distance in the time they took), they were heroic adventurers in an age when popular interest in records and feats of daring was huge. Byrd also attempted in 1927 to claim the Orteig Prize (worth $25,000 or about $345,000 today) for the first non-stop flight from the United States to France. A crash during a practice run put him out of contention and he was beaten by Charles Lindbergh, who completed the flight on May 21, 1927.

Lucky Lindy, as he was nicknamed, flew the *Spirit of St. Louis* solo across the Atlantic in 33½ hours, and set off a world-wide media frenzy. When he got home, he toured and gave speeches across the United States, Canada, Mexico and twelve other countries. President Calvin Coolidge presented him with the Congressional Medal of Honor in 1929.

This was also the beginning of the Golden Age of the dirigible. Dirigibles had been used for bombing and reconnaissance during the First World War, but they were vulnerable to attack. The large lifting capacity offered the possibility of passenger and high-end cargo transport. The *Graf Zeppelin*, one of the largest dirigibles ever built, took twenty-one days to circumnavigate the globe in 1929. Between 1930 and 1936, there were scheduled Zeppelin flights between Germany and Brazil. Dirigible travel was dealt a severe blow when the *Hindenburg* caught fire, killing thirteen passengers, twenty-two air crew and one ground crew. The event was highly publicized, since the landing and subsequent disaster were filmed by four newsreel crews, and radio reporter Herbert Morrison (1905–89) recorded his eye-witness account of the event.

Lindbergh's involvement with aviation continued through the inter-war years as he was invited by the German air force to inspect their new aircraft. Lindbergh became convinced that Germany would win any future war against France and Britain because German aircraft were so superior to anything the other countries had, and that the German military commanders understood the future of war. He also advised Congress that America should remain neutral in any future European war. In this way, Lindbergh became one of the first major public figures to comment on the role of technology for the wider world, linking technology to political decisions. The Lindbergh story took a tragic turn when the twenty-month old Charles Lindbergh Jr. was kidnapped and killed in 1932.

It is difficult to gauge how important such sensational events as Lindbergh's flight were to the popular belief in the power of technology, but in terms of making people aware of the changes to society that technology was creating, they were unmatched. People who had never seen an airplane in real life knew all about them from the extensive news coverage and articles in magazines such as *The English Mechanic and World of Science* (UK founded in 1865), *Popular Science Monthly* (USA 1872), *Popular Mechanics* (USA 1902) and *La Science et la Vie* (France 1913). These magazines promoted exactly the world that Lindbergh and the other adventurers were promising: the conquest of time and space in a world of mechanical marvels.

In the elation of the post-war years, the triumph of technology seemed everywhere in America, but the economy was being propped up by speculation and loose monetary policies. This led to the Great Depression, which began when the American stock market dropped on October 29, 1929. Although the roots of the depression predated the war, "Black Tuesday," as it would be called, started a wave of economic damage that spanned the globe. Many factors contributed to the stock market collapse, including speculation and buying stocks on margin.[2] Technology also played a role in creating the depression and the market panic. The industrial powers, particularly Britain, Germany and France, attempting to recover from the war, over-produced, but with little possibility of gaining markets or even maintaining existing ones because of the debts incurred during the war. The actual market for manufactured goods fell. This situation was made worse by monetary policies that prevented government intervention in the early stages of the decline, and reactionary trade barriers such as the imposition of high tariffs to protect domestic production after the market fell.

The world learned about the collapse of the US stock market very quickly, in part because of the speed at which stock information reached the markets. The idea for a stock market telegraph came from Edward A. Calahan in 1867, who worked for the American Telegraph Company. Thomas Edison's first really financially successful invention was his improved stock ticker, the "Universal Stock Printer." The device was essentially a dedicated telegraph line connected to a printer that recorded the trading of stocks. As sales of stock increased on October 29, the ticker was overloaded and lagged further and further behind the market's activities. This added to the panic, as investors could not follow the market and were forced to wait for information, or had their orders delayed, losing millions as stocks fell before the orders to sell could be processed.

❧ The Slide into War

The Great Depression reached its lowest point in 1933 and was easing by 1938, but political tensions were rising. Japan attacked China, and in Europe, Germany and Italy came under fascist control. From 1936 to 1939 the Spanish Civil War raged, and seemed like a test of ideologies, with communists, socialists and liberal democrats fighting conservative Nationalists and fascists. Germany supplied armor, aircraft and weapons and got an important military testing ground as they helped the Nationalists win the war. The technology of death had been increasingly mechanized, both by the application of industrial techniques to the production of war materials starting in the Industrial Revolution era, and by the end of the First World War by an increasing use of command and control systems that were built on concepts of industrial efficiency and modern communications. Although the First World War was global, in the sense that people from around the world participated in the fight and the war affected global commerce, most of the combat was confined to Europe and the Ottoman Empire. Of the approximately 70 million combatants, 60 million of them were Europeans. The Second World War would reach around the globe, with major battles on three continents involving more than 100 million soldiers from thirty countries.

The Great War had not resolved the underlying problems of natural resources and international relations that had sparked the earlier conflict. The economic situation helped to make populations susceptible to the simple solutions proposed by Hitler and other dictatorial leaders around the world. The move by Germany to prepare for another war was made possible by the belief among the German military leaders that a modern, technologically superior army could defeat the armies of the other European powers and gain what had not been gained in the earlier war: territory, natural resources, and political and cultural hegemony.

There is an old adage that the victors of one war plan to fight the next war in the same way, while the losers must find a new way to fight. The response of France to the threat of German rearming was to construct the Maginot Line, named after André Maginot, who was the Minister of Veteran Affairs, and later the Minister of War. The Maginot Line was a massive set of defensive positions running from Switzerland to Luxembourg and smaller defensive positions running from Luxembourg to the English Channel (Figure 10.1). It included hundreds of kilometers of trenches, pill boxes, gun emplacements and fortifications that were the equivalent of land-locked battleships complete with heavy guns, command centers, mess halls and barracks. The project cost over 3 billion francs.

The German high command had been removed at the end of the First World War, and the new officers, many of whom had fought in the trenches, were determined to avoid the stalemate and war of attrition of trench warfare. Ironically, one of the most significant sources for a new military model came from Major General J. F. C. Fuller (1878–1966), a decorated British officer who promoted the idea of mechanized warfare. Fuller's ideas were not well received in Britain since he criticized the British commanders and proposed to toss out the military doctrines being used by the British military. In 1926, he published *The Foundations of the Science of War*, outlining what he saw as the scientific principles of war. Although he was influenced by more than a hint of

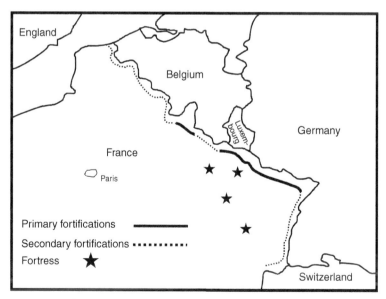

Figure 10.1 The Maginot Line.

mysticism (Fuller also wrote about yoga and occultist Aleister Crowley), his ideas were a virtual template for the German doctrine of Blitzkrieg or "lightning war." In the 1930s he joined the British fascist movement, and in 1939 he was a guest at Hitler's fiftieth birthday and observed the mechanized army of the Third Reich.

The German military adopted the doctrine of Blitzkrieg in large part because it was based on the technology of mechanized warfare. The attacker would use surprise, speed and combined forces to overwhelm the enemy. The plan was to strike so quickly that the enemy could not respond fast enough to reinforce the attacked point. By cutting through the front line at specific points rather than fighting along the line, the attacking force could cut off or capture communications and supply lines. Places of high enemy concentration such as fortifications were not to be attacked head on but to be isolated from supply and information, and could be more easily "mopped up" by forces following behind the first wave of attackers.

To achieve this, the German military instituted a program of integrated military operations that used airpower, artillery, armor and infantry together. The Luftwaffe was effectively a form of long-range artillery and provided support for armored units that moved quickly through the front line and into the rear of enemy positions. Mechanized artillery and infantry moved up to attack positions and hold territory. This had the effect of stretching out the depth of front lines from the few kilometers of the First World War to hundreds of kilometers, reflecting the effective range of bombers. Many of these principles were tested during the Spanish Civil War (1936–9), when Germany supplied tanks and bombers to the Nationalists.

At one level, the German strategy was brilliant. It really did use the technological capabilities of the various branches of the military to their best advantage. When the war started in 1939, the Germans scored quick victories over Poland, Belgium and France, which were overwhelmed by the speed of the attacks, largely unprepared to deal with armored assaults, and unable to counter German air superiority. The Blitzkrieg was more successful than the German military had dared to hope.

The success of the Blitzkrieg actually disguised some significant problems with German military doctrine. The high-technology solution was itself a trap. Keeping the highly mechanized military going required a very large infrastructure and huge quantities of resources, especially petroleum and munitions. The problem of fuel, lubricants, explosives and base stocks for industrial production (such as iron and nitric acid) became even more pressing when in 1941 Hitler broke the Molotov–Ribbentrop Pact and attacked the Soviet Union, opening a second front that had to be supplied.

Germany failed to force Britain out of the war, revealing that the Blitzkrieg had limits. Devastating as the air war was during the Battle of Britain, the expected invasion never happened. The German failure to develop long-range bombers left large sections of the British Isles free of attack and thus offered safe operational areas for the defenders. Despite major victories over Allied forces on every front, the German military failed to capture the oil fields of the Soviet Union or in the Middle East.

In addition to the technological trap of a highly mechanized military, the Blitzkrieg also had a functional problem. It was designed for aggressive attack, particularly against a static defender. It was not a strategy that led naturally to holding captured territory. Brutal oppression would keep most of the civilian population under control, but Germany lacked the manpower to effectively utilize the captured territory and had no political methods for dealing with subjugated populations. Even if the Second World War did not end up with trench warfare, it still became a war of attrition. A good example of this was the Battle of Kursk. After Germany lost the Battle of Stalingrad (1942–3), they pulled back and waited for resupply, and in particular the arrival of new Tiger and Panzer tanks that could take on the Soviet T-34 tanks. The Battle of Kursk during July and August 1943 was the largest tank battle in history, and pitted the technically sophisticated German tanks against the simpler but mass-produced Soviet T-34s. The German attack was blunted by defensive lines, and since the Soviet Army was willing to absorb huge losses, they simply wore down the Germans, forcing the military units to use up fuel, ammunition and food, and slowly reducing the number of tanks, trucks and aircraft. Although there were static defenses at Kursk, the Blitzkrieg failed in large part because the enemy as a whole was not static. Soviet tanks were constantly moving, and were given air cover by a determined, if technically less advanced, air force.

In hindsight, it is clear that by 1943 Germany could not expect to consolidate its victories. New aircraft, tanks and U-boats entered service, but once the industrial capacity of the United States started to resupply the Allies and American forces were mobilized, Germany was in trouble. Even before the American participation could be felt, in 1942 Germany began building its own version of the Maginot Line called the Atlantikwall to defend the western coast of Europe from invasion across the English Channel. Although fighting along the Atlantikwall was far greater than was experienced along the Maginot Line, in the end it suffered the same fate: surpassed, cut off and captured by a fast-moving, combined-forces attack. Sandwiched between Soviet forces in the east and Anglo-American forces in the west, the German military was crushed.

✿ Weapons of War

Historians of technology have debated the effect of war on innovation. The course of combat has definitely been influenced by new devices and tactics, but during a time of war it can be very risky to expend resources on experimental and untested weapons. In most cases, even if new weapons and systems are used during the course of a war, the perfection of such devices comes in the years following it. For example, aircraft were used during the First World War, but there were far more innovations in aeronautics between the wars than during that war.

The shift to technological innovation during a time of war started with the First World War, in large part because the war was so long that it was possible to create, test and bring into production new weapons such as mustard gas and tanks. One of the consequences of that war was the creation of an infrastructure for militarized science research in Germany, Britain, France and the United States. Thus, by 1939 many of the new weapons of the German military were technologically superior to those of its opponents, but the technology gap was not so great that simply having new weapons would give Germany victory. The weapons had to be combined with new tactics, and that proved to be decisive in the early days of the war. As the war dragged on, however, there was time for the Allies to create new weapons, tactics, and command and control systems such as radar and long-range bombers. Although Germany introduced a number of new "super" weapons during the war, lack of resources and workers (even using slave labor) limited production. In addition, many of Germany's greatest scientists had fled (and many would work for the Allies), were in jail or had been executed. Allied bombing destroyed infrastructure and the increasingly chaotic German high command added to production problems. A good example of this problem was the German Tiger II tank. No Allied tank came close to its power until near the end of the war, but less than 500 of them made it into service, while 40,000 American Sherman tanks rolled off the assembly line. Five specific technological innovations deserve special mention: codes, aircraft, radar, rockets and nuclear weapons. Each of these areas of military research played a significant role in the war and had significant impact on the world after the war.

Codes

At the heart of the German Blitzkrieg was information. Powerful as the German forces were, the Blitzkrieg required extreme coordination of groups moving at different speeds, coming from different locations and with different objectives. The only way to achieve this was by being directed into battle using a centralized command and control system that gave commanders the ability to follow events at a distance and give orders that would be carried out as soon as they were issued. Some parts of this system depended on telegraph and telephone systems, but radios were the only devices capable of maintaining contact with forces on the move through enemy territory.

The problem for all military forces was that information broadcast by radio could be intercepted by the enemy, and so a method of coding the information was necessary. The Germans used the Enigma machine, an electromechanical device that used a keyboard and a series of movable discs (called rotors) to encode and decode messages. Each time a key was pressed, a letter substitute was produced and the rotors would turn. Even if the same letter was entered twice, the output would be different. To decode a message, you would have to know the sequence of rotations necessary to turn the coded letter into the actual message.

In theory, the Enigma system was very secure, since there were nearly 159 quintillion (159×10^{18}) possible combinations, but Allied forces gained an advantage over the German military when a team of Polish mathematicians deciphered some early Enigma messages, starting in 1932. That information, along with some captured equipment and a good understanding of the procedures that German military units followed (for example, unit names would appear in every message, giving the code breakers a common clue), allowed Allied code breakers to decrypt tens of thousands of messages. The Allies also broke a number of Japanese codes.

The main center of Enigma code breaking was Station X, or Bletchley Park, an estate in Buckinghamshire, England. Codenamed "Ultra," the establishment brought together linguists, mathematicians and cryptanalysts. Among the code breakers was Alan Turing (1912–54), who would become one of the leading theorists and developers of computers.

In the later stages of the war, the German army modified the Enigma machines and the key codes changed more frequently, so that manual decoding became impossible. In response, the electrical engineer Tommy Flowers (1905–98) and his team from the Post Office Research Station were brought in to make an electronic decoding machine. The result was Colossus, one of the earliest digital electronic computers. The existence of the Colossus machines was kept secret until well after the war, and most of the wartime machines were ordered to be destroyed.[3]

Allied cipher machines were similar in principle to the Enigma machines, using electronics and physical rotors, but neither the British Type X nor the American SIGABA cipher machines were ever known to be broken by Axis code breakers. The problem with the Allied code machines was that they were big and complex, and could not be used anywhere they might fall into enemy control. In many cases a much less technological solution was to use "code talkers" or people who spoke rare languages such as Navajo, Assiniboine and Basque.

Aircraft

In the First World War, aircraft played a very small part in the war, despite a rather romantic view of the air war that pitted heroic figures like Billy Bishop against Manfred von Richthofen (better known as the Red Baron). German dirigibles bombed targets in Britain and Poland, but the damage was minimal and airships were easy targets for the much faster and more maneuverable airplanes. Aircraft, both fixed-wing and airships, played a more important role as scouts and spotters, working to counter naval attacks and spot U-boats at sea and observing troop movements and directing artillery on land.

Despite the limitations, many aircraft builders and military commanders believed that aircraft would play a growing role in future conflicts. Between the wars there were major developments in motors, control systems and armament, but the most significant innovation was the perfection of the mono-wing design. This made aircraft faster and more maneuverable. Combined with more powerful engines, the new aircraft could be built of heavier materials, lift more weight, travel faster and fly farther. Germany introduced the Messerschmitt Bf 109 in 1935 and the British Supermarine Spitfire began flying in 1936. Both pre-war aircraft would see significant use until replaced by newer designs.

Another major development was the introduction of strategic bombers. During the 1930s there was concern that strategic bombing (including the use of gas weapons) could destroy industry and terrorize civilian populations to the point that a country could not continue fighting. In Germany, the emphasis was on light or ground support bombers and medium bombers. The Junkers Ju 87, better known as the Stuka (a contraction of the German word for dive bomber, _Sturzkampfflugzeug_) and the Junkers Ju 88A were used in close ground support and tactical bombing. The Heinkel He 111, a twin-engine medium bomber, was used for strategic bombing. These aircraft worked very well in the first years of the war, helping German forces overwhelm the defenses of the countries it invaded. The Luftwaffe did build some heavy bombers, particularly the four-engine Heinkel He 177, but the Blitzkrieg doctrine emphasized combined arms where the air force operated in conjunction with ground forces and not as an independent combat unit. The ongoing demand for mass production of successful models limited the development of longer-range and heavier-load aircraft.

In contrast, Allied air forces introduced a number of long-range bombers such as the British Handley Page Halifax and Avro Lancaster and the American Boeing B-17 Flying Fortress and Boeing B-29 Superfortress. With these aircraft, Allied forces could strike far inside enemy territory. In fact, the bombers could fly so far that a new generation of fighters had to be created to provide defenses for the bombers. The most famous of the long-range fighters was the North American Aviation P-51 Mustang, which had a range of 2,755 km (1,650 miles). In comparison, the Messerschmitt Bf 109 only had a range of 850 km (528 miles) (Table 10.1).

Combat aircraft became increasingly complex over the course of the war. New electrical control systems replaced the simple wires and mechanical systems of the earlier

Table 10.1 Range of bombers.

Country	Aircraft	Range
UK	Halifax	3,000 km (1,860 miles)
UK	Lancaster	4,070 km (2,500 miles)
USA	B-17 Flying Fortress	3,220 km (2,000 miles)
USA	B-29 Superfortress	5,230 km (3,250 miles)
Germany	He 111	2,200 km (1,367 miles)
Germany	He 177	5,000 km (3,000 miles)

airplanes, radios and later radar were added, more instruments and navigational aids were installed that allowed for blind or instrument flight, and for bombers sophisticated bomb-aiming equipment was developed. In addition, the increased carrying capacity of large airplanes allowed for the introduction of paratroopers. Paratroopers were used by both the Allied and the Axis forces in the course of the war.

In the quest for faster aircraft, new propulsion systems were tried. Rocket-powered prototype aircraft were experimented with as early as 1928, while a working turbojet, the German Heinkel He 178, was flown in 1939. Italy, Britain and the United States had all flown jet prototypes by 1941, but the first production jet aircraft to enter service was the Messerschmitt Me 262. It appeared in 1944 and was the fastest production aircraft of the war, with a top speed of 900 km/h (559 mph). The late introduction of the Me 262 and its small numbers limited its impact on the air war, but its success in combat encouraged the development of jet aircraft by Allied air forces, and jet propulsion became the basis for modern fighters and bombers following the war.

The concept of air superiority became a major component of warfare because of the importance of air power during the Second World War. Following the war, new technologies for engines and avionics continued to be created, leading to ever more complex aircraft. Today, the most advanced combat aircraft cost millions of dollars to build and maintain. They also require massive computer power to operate, as systems and performance characteristics exceed human reaction times. Although the doctrine of air superiority was proven to be crucial in conventional warfare such as the 1967 war between Israel and Egypt, the Vietnam war and a series of conflicts in Afghanistan have shown that air superiority by itself does not guarantee military victory.

The development of new aircraft also had commercial applications. Commercial flights started after the First World War and by 1926 there was sufficient traffic that the United States introduced the Air Commerce Act, but it was largely after the Second World War that passenger and cargo flights became an important part of global commerce.

Radar

The discovery that radio waves could bounce off distant objects was recognized in 1904 by Christian Huelsmeyer (1881–1957). He created the telemobiloscope that used a spark gap to produce radio waves and then a pair of antennas to detect the reflected signal. He could get signals from 3 km (1.8 miles) away, but the system could not tell the distance to the object, only the general proximity. Huelsmeyer hoped that naval authorities would purchase his device to keep ships safe at night and in foggy weather.

In the 1920s and 1930s a number of groups worked on radar, including the US Naval Research Laboratory and the Compagnie Générale de Télégraphie Sans Fil in France, but the first successful use of radar as a wartime device was by the British Air Ministry. In 1934, Robert Watson-Watt (1892–1973)[4] was head of the Radio Research Station at Ditton Park. He was asked by the Air Ministry to comment on the possibility of using radio waves as a kind of death ray after claims made by German sources that Germany possessed such a device. Watson-Watt wrote back that such a device was highly unlikely, but radio waves could be used to detect aircraft. In 1935, Watson-Watt demonstrated that he could detect a bomber flying 13 km (8 miles) away. The Daventry Experiment convinced the government and military officials to develop radio detection, and by 1940 Watson-Watt and researchers from the Telecommunication Research Establishment had produced a working radar system. Known as RDF (radio detection finder) in Britain, in 1940 it was called radar (for RAdio Detecting And Ranging) by the *New York Times* and that name stuck.

The Air Ministry built a small number of radar stations known as Chain Home along the east and south coasts of England. The system would eventual include some sixty radar stations. While the radar was not perfect, it gave the British defenders an advantage over the German air force, allowing the Royal Air Force to concentrate its defenses by guiding fighters to the incoming bombers during and after the Battle of Britain.

German scientists also developed radar known as "Freya" for use during the war. It was technically better than the Chain Home system, but initially there were too few machines and too few trained operators to make them effective. As the war progressed, German radar improved, but it was never as effective as the Allied air forces quickly learned to jam the signals or exploit weaknesses in the German system.

Radar systems were also developed by other countries, including Japan, the United States, the Soviet Union and Canada. The sharing of British microwave technology, especially the disclosure of the cavity magnetron (a vacuum tube that generated microwaves for radar) by the Tizard Mission to the United States in 1940, opened the door to further improvements. After the war, radar became an integral part of command and control systems in both the military and civilian realms, controlling air traffic, guiding missiles and keeping track of ships at sea. As radar technology was made smaller and more compact, it even started showing up in the automobile industry, first in the hands of police catching people breaking the speed limits and then in automobiles themselves as part of collision warning systems, autonomous breaking controls and self-driving vehicles.

Rockets

The use of rockets in war dates back to the ancient Chinese, but, for the most part, rockets have played only a minor role in military weaponry, largely because artillery was easier to use and more accurate. The use of rockets in the Second World War did not really change that situation, but it did introduce new equipment that would make rockets important in the post-war years. On March 16, 1926 Robert Goddard (1882–1945) launched the first successful test rocket that used liquid propellant rather than a solid fuel like gunpowder. Goddard would continue to work on liquid-fueled rockets until 1941, but, in part because of a lack of support for equipment, his rockets reached a maximum of only about 2,700 m (9,000 ft), at a time when aircraft could fly higher than 15 km (10 miles).

In Germany, interest in rocketry focused mainly in the Verein für Raumschiffahrt (VfR) meaning "Spaceflight Society." A number of people, including Hermann Oberth (1894–1989), Walter Dornberger (1895–1980) and Wernher von Braun (1912–77), were associated with the VfR and later worked for the German military rocket program at Peenemünde, on Usedom island. Von Braun became the technical director of the Army Rocket Centre in 1937, but for most of the war the rocket program was a low priority. In December 1942, Hitler approved the development of the A-4 (Aggregat 4) rocket, later renamed V-2 (Vergeltungswaffe or Vengeance Weapon 2), but it was not until September 7, 1944 that the rocket was launched at England.

The V-2 was a liquid-fueled ballistic missile. It had a range of 320 km (200 miles), with a maximum altitude of 88 km (55 miles). It was an unstoppable weapon, traveling at more than 800 m/s (1,790 mph) on impact, and carrying a payload of 980 kg (2,200 lb). Over 3000 V-2s were launched, resulting in the death of more than 7,000 people. The rockets were constructed by slave labor, and more than 9,000 workers died building them, thus ironically killing more people by their manufacturing than their military use. The V-2 was a terror weapon, since it could not be targeted. It was simply pointed toward the general area of the target and launched. Terrifying as the sudden death represented by the V-2 was, it had no real effect on the war.

At the end of the war, the Peenemünde team was in danger of capture by the Soviets, so von Braun and most of his engineers and scientists fled across country to surrender to American forces. He and his team were recruited into the US Army and worked on a series of rocket projects, resulting in the creation of military missiles, including the Redstone and Jupiter-C ballistic missiles. In 1960, von Braun was transferred to NASA (National Aeronautics and Space Administration), becoming the director of the Marshall Space Flight Center in Huntsville, Alabama.

As rocket technology improved, the military utility of rockets became greater, particularly with the possibility of creating rockets powerful enough to carry nuclear weapons. The space age would really become part of public consciousness in 1957, with the launch of Sputnik I, the first artificial satellite, and Sputnik II which was large enough to carry the dog Laika into orbit. With a cargo capacity of about 500 kg (1,100 lb), Sputnik II was capable of lifting a nuclear weapon to orbit.

Nuclear Weapons

There is a popular belief that Albert Einstein was responsible for the birth of nuclear weapons. This is wrong. Although Einstein's 1905 equation $E = mc^2$ can tell you how much energy is in a given quantity of mass, it doesn't tell you how to get the energy out. The real origins of nuclear weapons came from two other sources, first the work of Otto Hahn, Lise Meitner and Otto Robert Frisch, and second Leo Szilard. From Hahn, Meitner and Frisch came the concept of nuclear fission, whereby a neutron ejected from the nucleus of an atom could smash into the nucleus of an atom of a heavy element such as uranium and break it apart, creating two lighter elements, and releasing energy and more neutrons. From Szilard came the idea of a nuclear chain reaction, in which one neutron striking the nucleus of an atom causes the release of two neutrons, and those two would release four, then eight, sixteen and so on. Together, these two discoveries made nuclear power and weapons theoretically possible, although the technical requirements to control such events in the real world were enormous.

In the end, three things were (and are still today) needed to manufacture nuclear weapons: (1) intellectual resources such as physicists and engineers, (2) an industrial base capable of producing the nuclear materials and bomb parts, and (3) access to large quantities of uranium.

Leo Szilard was concerned that Germany had the first two required elements for nuclear weapons, and when Belgium surrendered to Germany the Belgian Congo came under its control. In 1940, the Belgian Congo was the greatest producer of uranium in the world. Szilard, with the help of other scientists including Einstein, worked to persuade the American government of the danger of such weapons. In response, President Roosevelt authorized the Manhattan Project that would develop a nuclear weapon. Although much of the popular story of the Manhattan Project focuses on the work of the research scientists under the direction of Robert Oppenheimer, the entire project employed more than 130,000 people, from miners to plumbers. Most did not know they were working on a superweapon.

Once it was scientifically determined by Enrico Fermi and his team at the University of Chicago that controlled fission worked and a chain reaction could be sustained, the biggest industrial problem was getting enough enriched uranium and plutonium to actually make a bomb. In 1940, the entire world supply of uranium was measured in grams, but kilograms would be needed to make weapons. The technical challenge was enormous, since in naturally occurring uranium ore, only 0.7 percent is bomb-grade U_{235} and the rest is U_{238}. Several methods of refining (or enriching) the uranium were tried. In the end, a rather brute force method was employed, passing uranium hexafluoride gas through filters where the slightly lighter U_{235} would pass through faster. The partially enriched U_{235} would then be sent through a calutron that used electromagnets to complete the enrichment process. Plutonium, on the other hand, was not naturally occurring and had to be manufactured in a nuclear reactor by bombarding uranium with neutrons and essentially creating a heavier element out of a lighter element.

On July 16, 1945, the first nuclear weapon was exploded during the "Trinity" test at Alamogordo, New Mexico. Robert Oppenheimer, the scientific leader of the program, quoting the *Bhagavad Gita*, said "Now I am become Death, the destroyer of worlds." From a scientific point of view, the test was the final proof that nuclear weapons worked, but from a political point of view there was a problem. The Germans surrendered on May 7, 1945. Post-war debriefing of German scientists made it clear that although the idea of a nuclear bomb was well understood, there had only been limited efforts to investigate or create such a weapon in Germany. Since nuclear weapons had been developed to counter German science and technology, once Germany was out of the war many scientists believed the rationale for using them disappeared. The US military, on the other hand, regarded nuclear weapons as a new weapon in the arsenal.

The road to the use of nuclear weapons started on December 7, 1941, when the Japanese navy launched a surprise attack on Pearl Harbor, the American naval base in Hawaii. The attack was primarily an air assault, and was in the short term a tactical success, sinking or damaging fifteen major ships, destroying 188 aircraft and killing 2,402 military personnel. It was a strategic failure, however, as it brought the USA into the Pacific theater with full force. Even tactically, it was less successful than it might have been. Of the ships sunk at Pearl Harbor, only three were permanently lost, and none of the fleet's aircraft carriers was in port at the time of the attack.

After a long and bloody battle across the Pacific, Allied forces were on the verge of an invasion of Japan. US military planners believed that the Japanese military would fight to the death to defend the Japanese mainland. The invasion planners expected there could be as many as a million Allied casualties and many more Japanese killed and wounded. Rather than invade, it was decided to use nuclear weapons on Japan. The uranium bomb "Little Boy" was dropped on Hiroshima on August 6, 1945 and the plutonium bomb "Fat Man" on Nagasaki on August 9, 1945. On August 15, 1945 Japan surrendered unconditionally. The technology of death had reached a new level of destruction.

At the end of the Second World War, four countries had the three requirements to make nuclear weapons: USA, Britain, USSR and Canada. In 1949, the USSR set off its first nuclear explosion, sending a shock through the Western nations. This started an arms race that saw a host of nations acquire, or attempt to acquire, nuclear weapons and led to the United States and the USSR mass-producing newer and bigger weapons. The manufacturing of fission weapons largely ceased to be a scientific endeavor in the 1950s as the basic principles had all been worked out, but a new generation of even more destructive nuclear weapons based on fusion (known as hydrogen bombs, H-bombs or thermonuclear weapons) was created. The first fully thermonuclear device (it was too large and complex to be used as a weapon) was tested by the United States in 1952 and produced a blast that was 450 times more powerful than the nuclear weapons dropped on Japan (Table 10.2).

Modern warfare, transformed by the Blitzkrieg and increasingly controlled by commanders away from the front lines because of radio, became less about brute force

Table 10.2 Acquisition of nuclear weapons.

Country	Year acquired
USA	1945
USSR	1949
UK	1952
France	1960
China	1964
Israel	1966 (unacknowledged)
India	1974
Pakistan	1998
North Korea	2006

and more about information. Cracking codes and tracking the enemy by radar could turn the course of battle more than building a bigger gun. The front lines grew wider as aircraft could deliver weapons, soldiers and materials hundreds or even thousands of kilometers from their bases. Mechanized armies could travel farther in a day than nineteenth-century military forces could travel in a week.

Scientists, who had started to work on military projects in the First World War, became far more integrated into the military efforts of the Second World War. There were a number of unexpected consequences to the development of scientific warfare. In the victorious countries, the number of scientists and engineers absorbed by military programs, especially building nuclear weapons, aviation and missiles, was enormous. The best and brightest were recruited from colleges and universities into the military, governmental and industrial organizations, and into companies making faster jets, bigger bombers, nuclear weapons and all the related systems such as radar, radio communications, rockets and guidance systems. Japan, prevented by treaty from having a standing army or doing military research (at least not directly), directed its best and brightest engineers and scientists into consumer electronics, telecommunications and transportation.

War and industry, travel and communications combined to draw the world together in a way that had never been seen before. Technologies that had been developed to solve local problems such as the control of railway trains had grown to become gigantic networks that literally spanned the globe and were even starting to orbit the planet. Local concerns could be transformed into international events, and international events such as World Fairs and the Olympic Games could be attended by almost anyone and viewed by people in every corner of the world. In many ways, the technological world after the Second World War was just what the pioneers of the new technologies of the late nineteenth and early twentieth centuries had hoped to produce – at least in the industrialized countries – a world where people commanded the forces of technology to promote a civilization of comfort and wealth. War, transformed into a battle of advanced technologies, still cast its dark shadow on the globalization of technology that otherwise seemed to offer a bright future for everyone.

1 If the First World War was the first industrial war, why might we say that the Second World War was the first war of advanced technology?
2 Why would education be called the most powerful invisible technology ever created?
3 What is the paradox of industrial automation?

NOTES

1. There was a brief bulge in birthrates in Australia, Canada and the United States during the post-war Baby Boom, but it only lasted about a decade.
2. Buying stock on margin meant that the buyer payed only a fraction of the market value of a stock, borrowing the rest from the broker or a bank. It allowed a person to buy $100 worth of stock for as little as $10. If the stock went up to $200, the buyer could sell and gain a profit of $100 for an investment of $10 ($200 minus the $90 loan from the broker and the $10 initial investment). If the stock fell in value to $50, the buyer still owed $90 and had to pay the difference, effectively paying $100 ($10 plus the $90 loan) for a stock worth $50.
3. Part of the secrecy may have been due to the fact that the British promoted or sold Enigma-type machines to various countries after the war, and then could break the codes.
4. Robert Watson-Watt was a distant relative of steam pioneer James Watt.

FURTHER READING

By the middle of the twentieth century, technology had increasingly become a national issue that extended beyond the general promotion through patent laws. One of the biggest commitments was to public education as examined in a series of short documentary videos *A History of Education* (2006). A more direct look at education in the age of industrialization is Laurence Brockliss and Nicola Sheldon, *Mass Education and the Limits of State Building, c. 1870–1930* (2012). The Second World War looms large over this period and William Hard McNeill presents an interesting overview in *The Pursuit of Power: Technology, Armed Force, and Society Since A.D. 1000* (1984). Looking at a specific development, a nice succinct overview of the Manhattan Project is (despite its title) Cameron Reed's "From Treasury Vault to the Manhattan Project" (2011), while Sean Dash's *The Manhattan Project* (2002) provides digital images and L. R. Groves' autobiographical story *Now It Can Be Told: The Story of the Manhattan Project* (1962) presents a more human face even if it leaves out some material that was still classified when he was writing.

11 The Digital Age

1955	SS *Spyros Niarchos* starts move to supertankers *USS Nautilus*, first nuclear-powered submarine
1970	Boeing 747 enters service
1977	Apple Computers founded
1986	Chernobyl nuclear disaster
1989	Exxon Valdez supertanker oil spill
1992	Commercial development of the internet

The introduction of electricity was a major turning point in technological and world history. It changed everything, from communications to entertainment. The telegraph and the telephone depended on electricity and from them sprang a network of wires that spanned the globe. To make long-distance communications reliable, amplifiers were needed, and from the vacuum tubes that strengthened telephone signals a host of electronic devices could be created. This led to new kinds of devices such as radios and ultimately the computer, which was a machine capable of controlling other machines. It also made technology inaccessible since people could not directly observe the way electronic machines worked. Computers, combined with global transportation, linked almost every part of the world into a technological network.

The modern world has been fundamentally reshaped by the digital age as computers have taken control of the networks that make the globalized world possible. The term "computer" originally meant a person who carried out computations. Such people and teams of people existed, often unacknowledged, from ancient times to the mid-1970s, making calculations for everything from taxes to space flight. The utility of such services grew with the complexity of our devices and the degree of precision we demanded of our knowledge. For example, basic geometry and naked eye astronomy was all that was necessary to navigate the oceans in the time of Christopher Columbus, but to run a global empire the British needed the mathematical precision of star charts made with telescopes and the ability to track time that could only be achieved by the use of naval chronometers.

As a machine, the computer has reversed the usual direction of technological development by becoming more generalized over time rather than more specialized. This is not to say that computing devices have not been created to function for very specific tasks. The electronic device that opens and closes elevator doors is controlled by a computer, but the general machine, whether a desktop PC or a supercomputer in a laboratory, can be made to do many things. The computer is also the first machine designed specifically to control other machines and then networks of machines.

Digital devices ranging from industrial robots to smart phones now surround us, but their roots extend back to ancient times. The story of the creation of the digital age weaves together three main threads: the effort to embody mathematics in physical systems, the development of electricity as a commercial product, and the communication of information over long distances. Each of these developments was significant on its own, but by coming together they have transformed our relationship with technology, with each other and with the natural world.

✿ The Origins of Digital Devices

Digital devices are, at their most basic level, the translation of a mathematical concept into a machine. The beauty of mathematics is that it can be used to do so much, from computation such as accounting to describing and mapping nature (geometry, trigonometry, cartography and calculus). The machines we associate with the "digital age" are electrical, but they need not be, and in fact the original digital technology was mechanical. The desire to embody calculation in a physical form can in some sense be found in the tally sticks from prehistoric times, but the development of a reusable tool for calculation can be seen in early calculation tables, literally grids scribed on flat boards, stones and table tops. These have been identified in Asia, India, the ancient cities of the Fertile Crescent, the Roman Empire and the Inca world. These devices were used in much the same way as an abacus, using tokens or small stones to keep track of addition and subtraction.

Yet the mechanization of calculation went far beyond basic mathematics, as the Antikythera mechanism demonstrates (Figure 11.1). This device is the oldest mechanical calculator yet discovered, and is thought to date back to about 150–100 BCE.

Discovered in 1901 in an ancient shipwreck, its actual function was not determined until decades later when in 1971 historian of science Derek de Solla Price and physicist Charalampos Karakalos took gamma-ray and x-rays pictures revealing the inner workings that had corroded into a solid block. What they found was a complex collection of finely manufactured bronze gears, including a differential gear (a gear that transmitted rotation to two or more gears that could rotate over different distances). In 2005, the companies X-Tek and Hewlett-Packard joined forces with astronomers, mathematicians and archeologists to create a computer model of the Antikythera machine and decipher the writing on the parts. The research demonstrated that the device was a sophisticated celestial computer that could map the position of the known planets, predict eclipses, function as a calendar and even give the dates for the Olympiads. It was a brilliant embodiment of mathematics and astronomy in a machine. Its very perfection

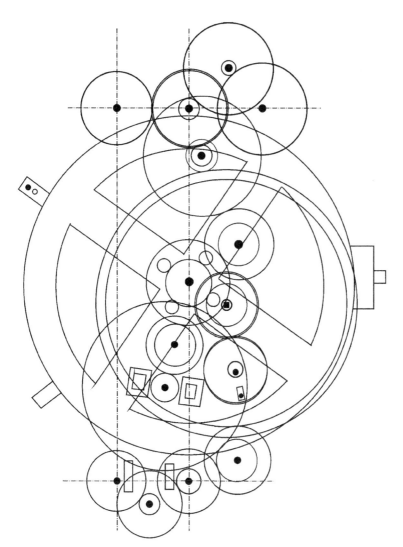

Figure 11.1 Antikythera mechanism.

suggests that it was not a singular achievement but the result of a long period of development, and proved that Greek mechanical skills were much more advanced than previously thought. The differential gear was to disappear from European devices until the eighteenth century.

A kind of super-abacus was developed by the Scottish mathematician John Napier (1550–1617), who introduced Napier's rods and later Napier's bones. These used flat rods and then four-sided sticks or "bones" inscribed with numbers that could be placed in a frame and moved to do multiplication and division, similar in action to a regular abacus, but using Indo-Arabic numbers directly (rather than counting beads) and allowing calculation of very large numbers. Napier went on to an even more important calculating aid with the creation of logarithms. A logarithm is an arithmetical function (an action preformed on a number) representing the power to which a fixed number must be raised to find a given number. Logarithms could be calculated for any range of numbers presented as a table. Multiplication of two large numbers, for example, was simply the addition of the logs of each number and looking up the result on the table. This allowed rapid calculation by generalizing the function so that it could be used for any set of numbers. Napier created logarithmic tables and they were expanded by others and contributed significantly to advances in mathematics, astronomy, engineering and physics. The concept of the logarithmic table was converted into a calculating device by William Oughtred (1575–1660) who realized that logarithms could be translated into a kind of graph, and that two logarithm graphs, if slid past each other to align parts of the graph, could be used to replicate mathematical functions. This created an analog computer.

The idea of logarithmic calculators was developed further by a number of people, but the slide rule in its modern form as a compact frame with a sliding central section was introduced around 1850 by Victor Mayer Amédée Mannheim (1831–1906), a French artillery lieutenant. It made sense that an artillery officer would be interested in calculations, since the artillery was the most technical branch of the military, and officers were expected to have a knowledge of chemistry, mathematics and the use of instruments. A rapid calculation might mean the difference between life and death. For generations, the slide rule was the constant tool of scientists and engineers until it was replaced by electronic calculators in the 1970s.

The French mathematician Blaise Pascal (1623–62) created a mechanical calculator made of gears, partly to help his father do accounting. Yet although he would go on to make as many as fifty calculating machines, they largely remained a novelty in mathematics. They had a tendency to jam, and people could do simple calculations faster in their head or on paper than they could input the numbers into the machine.

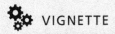 VIGNETTE

Charles Babbage

The greatest mechanical calculating devices of the Industrial Revolution were ones that were never quite completed. These were the Difference Engine and the Analytical Engine designed by Charles Babbage (1765–1871) (Figure 11.2). Babbage has been called the "Father of Computing," and has become an icon for the digital age. Although there is good historical reason to honor Babbage, his reputation in the field of computing must be tempered by the fact that he never actually produced working models of either of his calculating machines.

Babbage was born into a well-established family and was educated at the University of Cambridge. His mathematical skills were extremely good, and he was appointed the Lucasian Professor of Mathematics from 1828 to 1839, the same post held by Sir Isaac Newton in the seventeenth century and in the modern era by such mathematical giants as Paul Dirac and Stephen Hawking. Although Babbage felt that he had not achieved many of the things he had hoped, he was a powerful force for the organization of science and technological interests in industrial-era England. Along with friends and colleagues he founded the Analytical Society in 1812, he was elected a Fellow of the Royal Society in 1816 and he helped found the Royal Astronomical Society in 1820. After an attempt to reform the Royal Society failed, Babbage and other progressives who thought that science should be more open and have a wider audience founded the British Association for the Advancement of Science in 1832. In keeping with his interest in demonstrating the power and importance of mathematics, he also helped found the Statistical Society of London in 1864.

Babbage was one of the first people to consider seriously the consequences of industrialization. In 1833, he published *Economy of Manufactures and Machinery*, a detailed and sometimes critical evaluation of the state of British industry. He stated his aim in the Introduction: "The object of the present volume is to point out the effects and the advantages which arise from the use of tools and machines;– and to trace both the causes and the consequences of applying machinery to supersede the skill and power of the human arm" (Babbage 1835: 1).

Whether they superseded the human arm or the human brain, the advantages of machines were many: they never tired, got bored, lost focus, needed training or asked for more money. One of the great spurs for Babbage's effort to create a mechanical computer was that human computers (originally meaning people who did mathematical calculations) made so many mistakes, especially when doing repetitive jobs like compiling tables of logarithms. The Difference Engine would compute and print logarithms and trigonometric tables without error, using one of the most complex systems of precision gears ever conceived (Figure 11.3).

Babbage completed the designs for the Difference Engine and, using his many connections with government and the scientific community, obtained funding to construct it. The British government gave him grants because Babbage convinced them that accurate tables would aid navigation. Not satisfied with the limitations of the first

Figure 11.2 Charles Babbage.

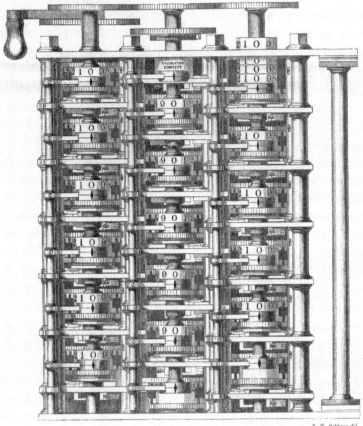

B. H. Babbage del.

Impression from a woodcut of a small portion of Mr. Babbage's Difference
Engine, No. 1, the property of Government, at present deposited in the
Museum of King's College, Somerset House.

It was commenced 1823.
This portion put together 1833.
The construction abandoned 1842.
This plate was printed, June, 1853.

Figure 11.3 Portion of the Difference Engine.

design, he created a second, larger Difference
Engine, but even as he was preparing to
construct the engine, his mind swept ahead
to an even greater project. The new engine,
which he called the Analytical Engine,
would not just compute a standard set of
mathematical functions but could be set to
compute any function. Where the Difference

Engine was a dedicated machine that could
do one thing – produce tables of polynomial
functions – the Analytical Engine could
be programmed to calculate a variety of
mathematical operations. Babbage borrowed
a control system that had been developed in
the textile industry – Jacquard cards. His idea
was that the punched cards would contain the

operations and the sequence that the machine would follow, while the machine would then perform the calculations and "store" the results (representing physical positions of gears) either to be used in the next step of the operation or to be read out as the answer.

Unfortunately, Babbage's ideas did not really get beyond the theoretical and design stages. Having failed to produce the Difference Engine despite several large grants from the British government, his pleas for funding for the more advanced machine were turned down. Babbage's story can teach a lesson about the conflict between perfection and completion. Although Babbage was well connected to the social networks of influence and money, his need to constantly revise his machine meant he could never demonstrate its utility.

❧ Computers I: Origins in the Electromechanical Age

The story of Babbage's engines has also been given a touch of romance and modern sensibilities by the participation of Ada Lovelace (Augusta Ada King, Countess of Lovelace, 1815–52), the daughter of Lord Byron. Lovelace had strong mathematical skills in an era when women were not generally considered part of the intellectual world. She was intrigued by Babbage's ideas for mechanical computers and offered to help. In a lengthy note appended to a translation she made of an Italian mathematical paper on the Analytical Engine in the 1840s, Lovelace presented a system for calculating Bernoulli numbers. This has been called the first computer program, and Lovelace the first computer programmer. While neither of these is exactly true, since Babbage had both conceived of the programming system and written "programs" prior to Lovelace's work, her contribution is still important because it shows that the concept of programming could be independent of the device.

Part of the reason that Babbage's work deserves more than a brief mention was that his machines would have worked. Pehr Georg Scheutz (1785–1873) built several Difference Engines based in part on Babbage's ideas. Ironically, he sold one to the British government in 1859. In 1991, the London Science Museum completed a working model of Babbage's Difference Engine number 2. It was composed of more than 25,000 individual parts, weighed 13,600 kg (29,982 lb), and was about 2.5 m (8 ft) tall. It worked as Babbage said it would.

With the mechanical skills made possible by the Industrial Revolution, basic calculating devices such as mechanical cash registers and adding machines were made possible, but they were not really much of an advance on the abacus and were often less useful for advanced mathematics than logarithm tables or slide rules. The next area of development was on the information side and it was driven by a need to keep track of people. One of the greatest migrations of history was taking place during the nineteenth century as people from around the world flooded into the United States. The government needed to keep track of the population, but the decennial census was growing into a massive operation as each bit of information for each person had to be recorded and tabulated by hand.

Herman Hollerith (1860–1929) was an engineering graduate from Columbia University. Hollerith was friendly with John Shaw Billings (1838–1913), formerly a surgeon in the US Army who had pioneered the use of medical statistics. Because of his background in statistics, he had been appointed as head of the Vital Statistics service of the US Census Bureau. Billings suggested that census information might be stored on cards similar to Jacquard cards. Hollerith took this idea and created both a system to punch cards and an electrical reader that used pins to complete a circuit so that the punched cards could be tabulated electromechanically rather than by hand. Hollerith, who was working at the Patent Office, patented his tabulating machines and used them to tabulate mortality statistics in Baltimore, New Jersey and in New York City in 1887. Hollerith built tabulators for the Census Bureau for use in the 1890 Census and his machines reduced the tabulating time to three months from the two years that the Census Bureau had expected when using the old method of tabulating by hand. The complete analysis of the 1890 Census took six years, two years shorter than the 1880 Census despite a significant increase in the population.

What made Hollerith's equipment so useful was not just his use of Jacquard's information storage, but the electrical equipment that was used to automatically tabulate the information stored on the cards. Using electromagnets and simple open/closed circuits, his equipment was on the cutting edge of electrical technology and would have been impossible to build only a few years earlier.

Despite the success of the tabulators, Hollerith resisted innovation and, like so many inventors, spent years in court defending his patents. Eventually he sold his company in 1912, although he continued to work as its chief consulting engineer until he retired in 1921. In 1924, the company was renamed International Business Machines Corporation. IBM, as it came to be known, would go on to be one of the most significant computer companies in the world, and it led the transition from the mechanical age of computing to the electronic.

Telephone I: The Electric Voice

With the technology of the telegraph established by the 1860s, a wide variety of inventors began to look for ways to transmit sound by electrical impulse. The story of telephony is full of controversy, with many claims for inventors from different countries being first to make a working telephone. Probably the earliest experimental model was made by Antonio Meucci (1808–89) in 1857. He filed a patent caveat (a document that described an invention and was a notice that a patent application would be filed at a later date) in 1871, but did not renew it, so others were free to patent their inventions.

In 1860, Johann Philipp Reis (1834–74) constructed a telephone in Germany, but was unable to interest either the scientific community or the telegraphy businesses in his invention.

In the United States, Alexander Graham Bell (1847–1922) and Elisha Gray (1835–1901) both built telephones along similar lines. Gray submitted a patent caveat in 1876, but Bell applied for a complete patent and was successful, beating Gray. The patent was controversial, since Bell's lawyer was owed money by the patent examiner Zenas Fisk Wilber. Wilber showed Bell's lawyer and Bell the caveat filed by Gray, and Bell later produced a prototype telephone that closely resembled Gray's design. In Bell's defense, he learned about Gray's design only after filing his own complete patent application, so he would likely have been granted the patent anyway.

The basic system of the early telephones was to transmit sound by converting it into electrical impulses with varying resistance. The first models used liquid to change the resistance and thereby the amount of electricity that was sent down the wire to an electromagnet that reproduced the sound by moving a diaphragm. Bell was assisted in his work by Thomas A. Watson (1854–1934). According to Bell, the first words spoken over their telephone were: "Mr. Watson – Come here – I want to see you."

The telephone was an important invention, but turning it into a system of mass communication required an additional invention. This was the telephone exchange, a device that would allow different telephone lines to be connected. The first successful exchange was invented by Tivadar Puskás (1844–93), who was working with Thomas Edison. The exchange was set up in Boston in 1877, and in 1879 Puskás set up the Paris telephone exchange. He also pioneered mass broadcasting over the telephone so that many people could listen to the same message. He used this to announce news and other programs in much the same way commercial radio was later used.

Controlling Electricity

The telegraph, electromechanical tabulators and the telephone were built around electromechanical devices that used very simple applications of electricity, either an on/off flow or a slight change in amplitude to vary the charge in an electromagnet. The move from electromechanical systems to electronic required a much greater ability to control electricity and manipulated the electromagnetic spectrum. To understand how this came about, we need to take a slight detour from the workbench of inventors into the physics laboratory of Heinrich Geissler (1814–79). Geissler was investigating the nature of electricity. He created a very good vacuum tube and ran an electrical charge through it from a cathode to an anode terminal, noting that there appeared an eerie bluish glow at one end of the tube near the anode. This suggested that something was being ejected from the cathode and traveling toward the anode. In England, William

Crookes (1832–1919) came up with the same basic piece of equipment. The "cathode rays," as they became known, fostered further research. On the continent, the general opinion was that the effect was a wave, while in Britain the phenomenon was generally considered to be due to a stream of particles. Although it was clearly electrical, and quickly shown to have a negative charge, it was a mystery as to what the glow meant. In 1894, physicist George Johnstone Stoney (1826–1911) named the mystery particle an electron, while in 1897 J. J. Thomson (1856–1940) measured the mass of the electron. Thomson then proposed a model for the atom that pictured a sphere of positive charge with negatively charged electrons embedded in it, sometimes called the "raisin-pudding" model. This model was challenged, first in 1903 by Hantaro Hagaoka (1865–1950) and his Saturn model, and then in 1911 by Ernest Rutherford (1871–1937) with his solar system model that placed a positive core at the center of the atom and orbiting electrons around it. Investigating the properties of the electron became possible because electrons could be produced and controlled by the vacuum tube. This led to the discovery of x-rays in 1895 by Wilhelm Konrad Roentgen (1845–1932) and, less directly, the discovery of radioactivity.

In the long run, the discovery of the electron and the model of the atom would get much more complicated as physicists looked further and further into the nucleus of the atom, but the technological impact of the devices created to study electricity were transformed into devices largely unexpected by the scientists. In 1906, Lee de Forest (1873–1961) placed a third terminal made up of a fine wire between the anode and the cathode inside a vacuum tube. De Forest found that a weak signal (pulses of electrons) could be amplified. The device, known variously as an Audion, a de Forest valve or a triode, allowed a very weak flow of electrons to be boosted in power. Although de Forest was denied a patent (his device was similar to a Fleming valve or diode), he partnered with the patent owner Guglielmo Marconi (1874–1937) to produce the new device. De Forest had been working on radio signals, but the Audion could be used for any electrical impulses such as those of telegraphs or telephones. By 1913, triodes were introduced as amplifiers for radio transmitters and on telephone lines, making long-distance calling possible.

✿ Telephone II: The Electronic Age

Using telephones for local calls was relatively simple, but to send the signal long distances was difficult because the electrical impulse faded due to the natural resistance in the copper wires. The maximum distance that a telephone signal could be sent was about 2,600 km (1,615 miles), equal to the distance between New York and Denver. The Audion boosted the signal along the wires. With this invention in place, Alexander Graham Bell ceremonially made the first transcontinental telephone call on January 25, 1915 from New York to San Francisco. In 1956, a consortium of the British Post Office, the American Telephone and Telegraph Company and the Canadian Overseas Telecommunications Corporation connected the first transatlantic cable between North America and Europe. Hawaii was connected to the continental United States in 1957 and from there to Japan and other Pacific locations in 1964.

Telephony had an enormous impact on global information flow. Businesses could communicate with customers and suppliers over vast distances and that led to greater centralization of business interests. News reporting, which had already been globalized by the telegraph, expanded further and increased the immediacy of news as reporters could dictate stories directly to copyists. The telephone made the centralization of control systems, whether in business, government or the military, even more effective. Ideas could be exchanged, questions asked and answered and orders given with the immediacy of face-to-face communication. Although private calls were expensive, they were more personal than a telegram. The telephone reproduced the familiar sound of the human voice with all its emotional nuances and erased the physical distance between callers.

✿ Radio

Joseph Henry, who had contributed so much to the development of electrical equipment, also contributed to the next stage of electrical communication when in 1832 he did experiments to detect electromagnetism at a distance and speculated on the existence of electromagnetic waves. This idea was worked out mathematically by James Clerk Maxwell (1831–79) in 1864, and described in his 1873 paper "A Dynamical Theory of the Electromagnetic Field," but it was Heinrich Rudolf Hertz (1857–94), working in Maxwell's laboratory, who demonstrated the transmission of electromagnetic signals. Hertz created a spark and demonstrated that it caused a spark to jump across a small gap in a nearby coil of copper wire. In one of the greatest underestimations of the utility of research, Hertz did not think there would be any practical application for his discovery.

Further notable investigations into "Hertz waves" were undertaken by Nikola Tesla, Jagadish Chandra Bose (1858–1937) and Thomas Edison. Each of these inventors worked on ways to control the creation of Hertz waves or on devices to detect Hertz waves. They shared a belief that Hertz waves, or as we know them today radio waves, could be used for wireless telegraphy. The Russian inventor Alexander Popov (1859–1905) constructed one of the earliest radio stations, setting up a transmitter on Hogland Island off the coast of Russia in 1900 that successfully transmitted and received signals from a battleship 40 km (25 miles) away.

Voice transmission was first accomplished by Reginald Fessenden (1866–1932) in 1900 when he was working for the United States Weather Bureau. The test of voice transmission took place on Cobb Island, Maryland over a distance of 1.6 km (1 mile). Fessenden got into a dispute over patent rights with his superiors, so continued his work at General Electric. Fessenden, with help from engineers and scientists at GE, created high-frequency alternator-transmitters that would allow sound to be transmitted by amplitude modulation. On December 24, 1906 Fessenden broadcast a short program from Brant Rock, Massachusetts that included him playing the violin and reading a passage from the Bible.

Important as Fessenden's work was, it was Guglielmo Marconi who made the greatest strides in turning radio into a commercial product. Marconi read the work of Tesla and Hertz, and in 1895 made a transmitter that could broadcast about 1.6 km (1 mile). In 1901, he claimed to have received a transatlantic signal, but there is some dispute over whether his apparatus could have received such a signal. It is certain, however, that in 1902 Marconi transmitted signals from Glace Bay, Nova Scotia to Ireland. In 1904, Marconi began commercial broadcasting (by Morse code) of news to ships at sea, and offering radio-telegraph services. In 1909, Marconi shared the Nobel Prize in Physics with Karl Ferdinand Braun for his work on radio. His company began to manufacture radio apparatus, and eventually began sound transmissions, establishing radio broadcasts in Britain starting in 1920.

The introduction of tube technology also transformed radio. With this technology, Edwin H. Armstrong (1890–1954) patented a new form of radio transmission based on frequency modulation (FM) rather than amplitude moderation. FM helped minimize atmospheric static and interference from electrical equipment. The first FM station, W1XOJ near Boston, Massachusetts, began broadcasting in 1937. A further advantage to FM broadcasting was the ability to transmit a stereo signal using two carrier waves known as L+R (left and right) main channel signal and L–R signal. This allowed sound to be recorded with two microphones (representing the difference between what the right and left ear would hear) and more realistically reproduced the sound using left and right speakers. This system became commercially available in the 1960s.

✿ Television

Another of the great branches of the use of the cathode ray tube was the creation of the television. Early electromechanical systems for the transmission of images appeared at the nineteenth century, particularly with the work of Arthur Korn (1870–1945) and Paul Gottlieb Nipkow (1860–1940), but did not really produce a useful product. The idea of sending images using electrical signals was investigated by so many people that the idea was widely known and the term "television" predated the first demonstration of a working television by a quarter of a century. Constantin Dmitrievich Perskyi (1854–1906) used the term "television" at the First International Congress of Electricity at the International World Fair in Paris in 1900, but the first public demonstration was presented in 1925 by John Logie Baird (1888–1946) at Selfridge's department store in London. Baird sent a television signal from London to New York in 1928 using selenium photoelectrical tubes to trigger an electrical signal that was converted into radio waves. The signal was converted back to light using a neon lamp.

Fully electronic television systems or components were introduced by a number of inventors, including Kálmán Tihanyi (1897–1947), Philo Farnsworth (1906–71) and Vladimir Zworykin (1888–1982). All of these developments built on the ability of new cameras that used photosensitive vacuum tubes to translate light into electrical impulses that could then be converted back into light by the receiving television. The year 1936 was key for television, with the BBC in Britain demonstrating a television broadcast system and the Heimann company in Germany broadcasting at the Berlin Olympic Games.

By 1940, commercial televisions were starting to be manufactured, but the Second World War interrupted production until after the war. In the industrial world there was an explosion of television consumption. In the United States, 9 percent of households had a television in 1950, but by 1965 the percentage had reached 92 percent. Although television did not wipe out commercial radio, it became a major conduit for news and entertainment. In 1960, the first US presidential debate took place between John F. Kennedy and Richard M. Nixon and was televised nationwide. It became popular belief that Kennedy won the first television debate, although there is no real evidence of this. What is true is that television transformed the way politicians communicated with the electorate.

✿ Computer II: From Electromechanical to Tubes

Devices created to use and control electricity and the telephone became available for other applications. In particular, the triode that was initially used as part of the amplification of signals in radio and telephones could control whether a charge was transmitted or not, oscillating very quickly from on to off. That meant the tube would be in a particular state, either charged or uncharged. In 1918, this capacity was utilized by the physicists William Eccles (1875–1966) and Frank Wilfred Jordan (1881–1941), when they designed the "flip-flop" circuit. This meant that binary states (on/off, zero/one) could be represented and controlled by electrical signals. By 1938, a number of people were working on the idea of calculating machines based on electronic signals, such as Alan Turing (1912–54) and John von Neumann (1903–57). The importance of the mathematics behind computers would be put to practical use during the course of the

Second World War, as Allied and Axis cryptologists worked to break each other's codes. Because of the complexity of modern codes and the volume of messages intercepted, first electromechanical and then electronic equipment was needed to do codebreaking, something that had been done manually before the war (Table 11.1).

Turing reasoned that any mathematical problem that could be solved could be calculated with a very simple machine. This was conceived as an abstract "machine" that used a table of rules. The Turing machine was not a specific physical machine, but rather an idea about doing mathematics that could be embodied in a physical system. The very simplest Turing machine could be conceived of as a long strip of paper divided into cells and a ladybug that could follow a simple set of instructions, such as "move forward x number of spaces," that could be written in each cell. By simply moving and following the instructions the problem of 2 + 2 = ? could be solved by the commands "move forward two spaces, move forward two spaces and say the number of the cell." The ladybug could solve complex problems if given enough time. Electronic computers are not Turing machines, but the theoretical principles of tables of instructions and conceptions of calculation that could be embodied by a physical machine helped to make electronic computers possible. What electronics offered was the ability to use the fastest information transfer system possible – the speed of electricity – to carry out the computational instructions.

At Harvard, in 1945, another team of scientists were working on a more general electronic computer. John Presper Eckert (1919–95) and John W. Mauchly (1907–80) and their assistants were building ENIAC, the Electronic Numerical Integrator and Calculator. It could be programmed and it could store information in electronic form. The war in the Pacific contributed to the development of technology on the part of the Allies, particularly leading to the creation of ENIAC, one of the first programmable electronic computers. ENIAC was paid for by the US Army, Ordnance Corps and created for the US Army's Ballistic Research Laboratory. It was to be used to create firing tables for artillery, but because it did not become fully operational until 1946 it was too late for war work. Although based at the Ballistics Research Laboratory, at the Aberdeen Proving Ground, Maryland, ENIAC came to the attention of John von Neumann, who was working on the hydrogen bomb at the Los Alamos National Laboratory. Scientists needed to calculate the distance neutrons would travel through different materials before they would collide with the nucleus of an atom of that material. Huge teams of people called "computers" were needed to carry out the repetitive calculations to solve this problem. Von Neumann recognized that ENIAC could more quickly and accurately do those calculations. Thus, the first major use of ENIAC was part of the Cold War arms race between the United States and the Soviet Union.

ENIAC was a major breakthrough in electronic computing, but it was not without problems. It contained around 18,000 vacuum tubes and tens of thousands of other electrical components including two 20-horsepower blowers to keep the electronics

Table 11.1 Pre-transistor computers.

Date	Computer	Maker
1940	Complex Number Calculator	Bell Telephone
1941	Z3	Konrad Zuse
	Bombe decryption machine	British Intelligence
1942	Atanasoff-Berry Computer	John Vincent Atanasoff at Iowa State College
1943	Project Whirlwind flight simulator using an analog computer	Massachusetts Institute of Technology
	Relay Interpolator	Bell Laboratories
1944	Harvard Mark-1	Howard Aiken at Harvard University
	Colossus (operational)	Tommy Flowers
1946	ENIAC	John Mauchly and J. Presper Eckert
1948	Selective Sequence Electronic Calculator	Wallace Eckert at IBM
1949	EDVAC	Maurice Wilkes at Cambridge University
	Manchester Mark I	Frederick Williams and Tom Kilburn at Manchester University
1950	ERA 1101	Engineering Research Associates
	SEAC (Standard Eastern Automatic Computer)	US National Bureau of Standards
	SWAC (Standard Western Automatic Computer)	US National Bureau of Standards
	Pilot ACE	UK National Physical Laboratory
1951	Whirlwind	Jay Forrester at MIT
	Lyons Electronic Office	Lyons Tea Company
	UNIVAC I	Remington Rand for US Census Bureau
1952	IAS computer (copied as MANIAC, ILLIAC, SILLIAC and others)	John von Neumann at the Institute for Advanced Studies

cool. It required constant maintenance while running. ENIAC also consumed around 160 kilowatts of power per hour for cooling and operations.

Even before ENIAC was built, plans were being made to construct a more sophisticated computer. The result was EDVAC (Electronic Discrete Variable Automatic Computer), completed in 1949. John von Neumann, a brilliant mathematician who consulted on the EDVAC project, wrote the First Draft of the report on EDVAC in

1945, making the basic ideas for electronic computing available to anyone with the technical knowledge to understand the report. Although the report was supposed to be confidential, it was copied and circulated as far away as Britain. Tommy Flowers, lead engineer of the British code breaking machine Colossus, offered suggestions for circuit designs. There was acrimony about von Neumann's report, since it damaged the possibility of patents for the electronics being developed, and, because it was published under von Neumann's name, credit for ideas created by other scientists were not acknowledged. Despite the controversy, the long-term impact of the First Draft was significant, influencing the design of computers.

ENIAC was followed by a number of other computers, but they were generally individually made and were more scientific instruments than general purpose machines aimed as some commercial market. They also suffered from the problems that all tube technology was subject to, namely using a great deal of power, producing large quantities of heat and tubes failing on a regular basis. The biggest commercial users of tube technology following the war were the telecommunications companies such as Bell Telephone, which used amplifier tubes to maintain signal strength on telephone lines. Whole armies of technicians were needed to monitor and replace tubes. Calls were interrupted or lines were unavailable when tubes failed. While mass production of tubes had reduced their cost, the maintenance issues were still considerable.

⚙ Solid-State Electronics

The physicist William Shockley (1910–89) had been working on the electrical properties of crystals such as silicon before the war, hoping to produce a solid-state version of the de Forest Audion tube. It had been discovered that crystals that contained traces of germanium would transmit a flow of electricity. After the war, Shockley returned to work at Bell Laboratories and his crystal studies. Shockley understood the physics, but he could not get solid-state materials to work. He got his two assistants, John Bardeen (1908–91) and Walter Brattain (1902–87), to work on the problem and in December 1948 they succeeded in creating the first solid-state amplifier. Shockley, Bardeen and Brattain would share the Nobel Prize in Physics in 1956 for their invention. The device was called a "transistor" by their colleague John R. Pierce (1910–2002) as a contraction of its property of "transresistance." The team showed their experimental work to Bell executives, who did not fully comprehend what they were seeing.[1] The prototype device, made of chunks of plastic, some wires, a strip of foil and a bit of crystal, all held in place by a bent paperclip, could be used as an amplifier, making tube technology obsolete, but what the Bell executives did not understand was that the transistor could be used for a huge range of things, not just amplification.

The transistor found a use in hearing aids, but its first major commercial success was when a small Japanese company, Tokyo Telecommunications Engineering Corporation, put transistors into portable radios small enough to fit into a shirt pocket. This was not the first transistor radio – early radios had been manufactured by the Regency Company and Texas Instruments – but the Sony TR-72 released in 1955 cracked the international market. In 1958, the company changed its name to Sonī Kabushiki Kaisha, or more commonly Sony Corporation. In the various models, by 1968 more than 5 million Sony transistor radios had been sold in the USA alone. The miniature radio hit the market at just the right time to take advantage of the Baby Boom generation who had disposable income, were fascinated by new things and were eager to be entertained.

❀ Telephone III: Combining Radio and Telephone

The idea of a mobile telephone was as old as the idea of the telephone, but until the invention of radio it was not technically possible. Station-to-station two-way radio communication had begun with Marconi, but portable two-way systems started to appear in the late 1930s. Donald Hings (1907–2004) created a two-way system for bush pilots in 1937, and Alfred (Irving) J. Gross (1918–2000) developed a number of designs between 1938 and 1941. Gross would go on to work on CB radio (Citizens Band) and the telephone pager. The first commercial products were developed for the US military. A team at the Galvin Manufacturing Company (later Motorola) produced a backpack model that went into US military use in 1940 and a hand-held model originally called the "handie talkie" in 1941.[2] Similar devices were being built by British, German and French inventors.

In many ways, the combination of radio and telephone was an invention looking for a market. The wartime radios were big and heavy, and used tube technology, but with the introduction of solid-state electronics they could be made more compact and much lighter. The first commercially available system was the MTA (Mobile Telephone system A) released by the Ericsson company in 1956 in Sweden. A specialty item and an inconvenient 40 kg (90 lb) in weight, it never had more than 600 customers. Finland had the ARP network in 1971, while the Motorola CynaTAC phone was premiered in 1973 in New York, but it was not until miniaturization and agreements about the radio frequencies that could be used for mobile telephones were reached in the 1990s that mobile phones began to be more than luxury items for the rich. The so-called G2 (second generation) phones were much lighter and started to add other services such as text messages to supplement the voice system. The first person-to-person text message was sent in Finland in 1993. By 2007, mobile telephones had, according to their producers, reached the fourth generation (4G) and were in effect hand-held computers with the capability of constant connection with the radio frequency network and seamlessly (as far as users are aware) integrated into the digital systems of existing telephone landlines and the internet. Although the development of the mobile telephone did not depend on the internet (which will be looked at presently), they were devices that depended on such closely related components that the integration of mass communications and computer systems was a logical step.

✿ Transportation

In the nineteenth century, canals and then railways transformed transportation and helped spread industrialization. By lowering the cost of moving materials and people, industrial centers could be made bigger and the hinterland supplying the raw materials was expanded. Today, transportation is an integrated global system that combines water, air and land transport. For much of the world, this system is so well integrated that there is no particular business distinction in the mode of transport. Getting goods and people from place to place has also created new jobs in transport logistics and supply chain management.

In the twentieth century, innovations in naval technology meant that the advances in ocean transportation continued. Radar and radios became common on ocean-going vessels after the Second World War, making navigation safer and faster. Although oil tankers had existed from the end of the nineteenth century, the era of the supertanker began in 1956. In that year, Israel, Britain and France invaded Egypt to gain control of the strategically important Suez Canal after it was nationalized by the Egyptian government. The Suez Crisis closed the Suez Canal that linked the Middle East, Africa and Asia to Europe and the Atlantic region. Until that time, oil tankers had been limited to the size of ship that could safely pass through the canal, but without the Suez Canal route tankers had to travel longer distances in the open sea. In 1955, the world's largest tanker was the SS *Spyros Niarchos*, with a 47,500 DWT or dead weight tonnage (total weight of ship and cargo). The Suez invasion failed when other Western countries and the USSR condemned the attack, but the fear that the canal route could be cut off at any time led to a change in design philosophy for tankers. In 1958, the SS *Universe Apollo* was launched. It had a DWT of 126,850. By the 1970s, supertankers were being built that had 250,000 DWT. Currently, the largest tankers are TI-class ships launched starting in 2002. They have a 441,500 DWT and can carry more than 3 million barrels of crude oil at a speed of 16 knots (30 km/h or 19 mph).

The economy of scale for ocean transport means that larger ships are more profitable, but there are both engineering and practical limits to the size of ships. Most ports cannot accommodate the giant tankers and they require specialized equipment for loading and unloading. They also pose serious environmental dangers. Although there have been numerous spills from tankers, the most famous was the *Exxon Valdez* disaster. On March 24, 1989, the *Exxon Valdez* (214,800 DWT) hit a reef off the coast of Alaska and lost 260,000 barrels of crude oil. Oil washed up on more than 2,000 km (1,300 miles) of coastline, killing marine life and shore animals. The spill led to increased regulations for tankers in American waters, but tanker transport traffic has increased, raising the possibility of new spills. An even bigger disaster, the Deepwater Horizon spill in the Gulf of Mexico on April 20, 2010, which released an estimated 5 million barrels of oil, highlighted the continuing danger of oil spills. The world's seemingly insatiable appetite for oil makes this technology one of the hardest to control and regulate.

Tankers also contributed to another ocean transport revolution. The first modern container ship was the *SS Ideal X*, built out of a converted oil tanker in 1956. Although the idea of shipping goods in boxes or large cases was very old, Malcom McLean (1913–2001), the owner of a trucking company, figured out a way to "containerize" goods so they could be moved by truck and then transferred to ships. This was the beginning of intermodal transport, where steel shipping containers could be moved by truck, train or sea. Containers come in a variety of sizes, but the two most common are the 20-foot and the 40-foot. The 40-foot works well with semi-trailer trucks. A 40-foot container is capable of carrying up to 26,500 kg (58,400 lb) of cargo.

Container ships were not well accepted at first because the specialized cranes and handling equipment were expensive, and longshoremen feared the loss of jobs that would follow. The economic advantages and flexibility of container transport eventual made them the main choice for global trade. The largest container ships can carry more than 14,000 containers, while smaller "feeder" ships carrying up to 3,000 containers can deliver cargo to smaller ports. Some container ships even have cranes so cargo can be delivered to places that lack container systems. Not including bulk transport (things like coal, grain or minerals), 90 percent of global cargo is now transported by container. The average price to ship a 20-foot container from Asia to Europe is around $1,500, or about $14 a tonne.[3] This has contributed to the globalization of manufacturing as transport costs declined sharply due to new methods of handling cargo and ceased to be a major consideration for the price of manufactured goods.

The vast increase in global demand for transportation has also filled the skies with airplanes. The aviation technology that developed during the Second World War, including the turboprop, pressurization, radar, radio and the jet engine, opened the door for the development of much larger and more efficient air transport in the post-war years. From 1952, jet airliners began to enter service for commercial transport, starting with the de Havilland Comet and soon followed by the Sud Aviation Caravelle (France), Tupolev Tu-104 (Soviet Union), Boeing 707 (USA), Douglas DC-8 (USA) and the Convair 880 (USA). These airplanes were used for cargo and for passengers, although the passenger market was limited because it was expensive.

A major shift in air travel happened when the Boeing 747 was introduced in 1970. Nicknamed the "Jumbo jet," the most common 747 configuration could carry 400 passengers at a cruising speed of 920 km/h (570 mph) (Figure 11.4). As a transport, it

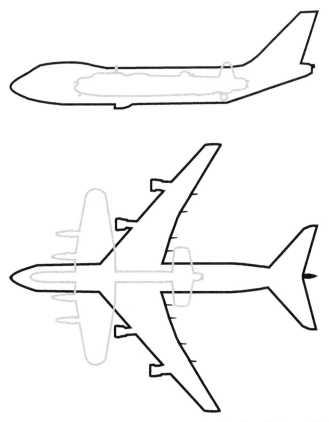

Figure 11.4 Comparing an Avro Lancaster bomber from the Second World War with a Boeing 747 from 1970.

could carry 140 tonnes (154 tons) of cargo.[4] This revolutionized air travel and made high-speed cargo transport affordable for a larger range of products. In a matter of years, air travel became common and the cost of travel dropped to the point where it was available to millions of people. Today there are between 9,000–10,000 commercial aircraft in the sky at any moment, and between 600,000 and 1 million passengers in flight. Between 1974 and 2015, global passenger traffic rose from 400 million passengers annually to almost 4 billion passengers. Impressive as the passenger statistics are, they are surpassed by the growth in air cargo transport. In 1974, 16,955 million tons of cargo were transported by air. In 2015, 195,162 million tons were transported and the delivery company FedEx was operating the fifth largest fleet of commercial aircraft in the world.

Transportation has also become so reliable that many industries now operate using just-in-time manufacturing (also known as continuous-flow or lean manufacturing). In this system, raw materials are delivered for processing as they are needed rather than factories having large stockpiles on hand. In turn, the component parts for manufactured goods are sent to the assembly plants in a continuous stream rather than as large shipments. An automobile factory may only have enough steering wheels on hand for one day of work, but a new load of steering wheels will arrive before the next day. This eliminates most of the storage space required for inventory and reduces handling (packing, unpacking, storing and getting materials out of storage). In turn, this allows for greater "on demand" manufacturing in which the finished product is created after it is ordered rather than a certain quantity being built based on an estimate of the number that will be sold.

The changes in transportation have changed our perception of time, distance, relationships and the seasons. Fresh tropical fruit is available in the middle of winter. Traveling to any place on the globe for a holiday is affordable for millions of people, and relocating to a distant country for work has become commonplace. In military terms, every place on the planet can be reached by military forces within hours. Transportation has transcended the world order established during and following the Industrial Revolution that placed manufacturing in the comparatively rich countries of Europe and North America, and transferred it to low labor cost regions such as Asia and South America. Unions, which grew up to represent the interests of workers during the eighteenth and nineteenth centuries in the industrialized countries, were circumvented by corporations that could outsource work because low labor costs and low transportation costs offered great profits. There are ongoing arguments amongst economists and people concerned about global labor equity about whether globalization has been a good thing or not. On one hand, there has been a growing middle class in China and India, but there has also been the creation of "rust belt" regions where heavy industry has been abandoned in the United States and Britain.

Cheap transport has also contributed to the incredible accumulation of "stuff." Today, the cost of bringing goods to market has declined so much, relative to the cost even a century ago, that the world is filling up with manufactured goods. The sheer number of physical objects that people own, particularly in the industrialized world, has grown because of the decline in the cost of production and the low cost of transport. All of those objects use resources and most will one day end up as trash. Ironically, rich countries can often afford to send their garbage to poor countries, using the inexpensive transport ships. The problem is so serious that the United Nations Environment Program issued a report entitled Waste Crimes (2015), looking at the international trade in hazardous waste and electronics.

⚙ Robots

One of the most significant uses of digital devices was the creation of a new generation of industrial robots. The idea of robots or autonomous mechanical devices (usually but not always in human form) extends back into antiquity, with stories about golems, living statues and clockwork machines. Hero of Alexandria (10–70 CE) designed a steam-driven bird, al-Jazari (1136–1206) made mechanical figures, Leonardo da Vinci (1452–1519) drew plans for a mechanical knight, while Hisashige Tanaka (1799–1881) built tea-serving dolls. The term "robot" was coined by the Czech artist Josef Čapek (1887–1945), and then used in his brother Karel Čapek's play *R.U.R.* (Rossum's Universal Robots) in 1920. These devices, really more entertainments than tools, nonetheless served in working out many of the basic mechanical aspects of robots. The first digital and programmable robot was invented by George Devol (1912–2011) and in 1961 was purchased by General Motors. It was used to stack hot pieces of metal. In 2018 there were more than 4 million programmable industrial robots worldwide.

Experiments with autonomous robots, devices designed to move independently and interact with a changing environment, began in 1948 when William Grey Walter (1910–77) created Elmer and Elsie as part of his work on neurology. Robots that play table tennis, look for landmines or imitate humans (often referred to as "androids") or are autonomous vehicles are all still at the developmental stage. There is a great deal of interest in military robotics, autonomous vehicles including driverless cars, and surgical robots, all of which offer the possibility of greater precision and safety. With the interest in robotics has come concern about the potential danger of autonomous devices – a frequent plot device in popular media from Frankenstein's monster to Robocop. Social concerns have also been raised over the possibility of using robots as companions for the elderly, the disabled and children. When the First International Congress on Love and Sex with Robots was held in 2014, it drew international media attention despite being strictly an academic conference.

Computers III: Solid-State and Personal

At about the same time as the Sony radio was hitting the market, transistors were being used in a computer. In 1953, Richard Grimsdale at the University of Manchester turned on the transistor-based computer simply called the Transistor Computer. It was an experimental platform and was followed by a more powerful version called Atlas in 1962. With the new technology, computers became more commercially viable, and one of the first successes was the IBM 1401. IBM built more than 100,000 of these between 1960 and 1964.

In 1958, the transistor was joined by other electronic components in the integrated circuit. Jack St. Clair Kilby (1923–2005), working at Texas Instruments, figured out a way to use solid-state materials to take on more functions than simply amplification of an electrical signal. In particular, logic gates used switching circuit theory to perform Boolean algebra, while flip-flop circuits would store binary data. One piece of binary data (a one or zero representing either an on or off state) was called a bit, with eight bits commonly called a byte. These functions had been possible with vacuum tubes, but solid-state versions could be made microscopic, eventually allowing millions of circuits to be built into a single chip. The same idea was independently discovered by Robert Norton Noyce (1927–90). Noyce was the co-founder of Fairchild Semiconductor, and went on to found Intel with Gordon E. Moore (b. 1929) in 1968. Moore is also famous for the idea that the number of transistors in an integrated circuit would double approximately every two years (see Moore 1965).[5] He stated this not as a law of nature, but as an observation about the electronics business.

The appearance of more and more computing power in a smaller and smaller space made it economical to use integrated circuits for a vast array of control mechanisms, from car engines to automated doors. Combined with a growing number of devices that were designed or redesigned to send a signal to a computer, such as pressure pads, infrared and laser scanners, weight scales, linear measuring instruments, thermometers and counting devices, there was almost nothing in the industrial world that could not be recorded and controlled by computers.

While digital systems made industrial robots possible, it was the introduction of small computers to business and later to individuals that had the most noticeable effect on society. These microcomputers introduced in 1975 such as the Altair 8800 and the IMSAI 8080 were inexpensive enough for people to buy, although they were not really designed for the average consumer. More consumer-friendly computers such as the Apple II and the Commodore PET appeared around 1977. Only a handful of computer enthusiasts thought that individuals would want or could find a use for computers, so the personal computer market was untapped. Apple Computers, founded in 1977 by Steve Jobs (b. 1955), Steve Wozniak (b. 1950) and Ronald Wayne (b. 1934), started in a garage and went on to be one of the most powerful computer and telecommunications companies in the world on the basis of bringing computers to individual consumers. By challenging the power of IBM, the hope was to defeat Big Brother and bring computing to the people. And make money at the same time.

When IBM entered the personal computer market, it set out to dominate it just as it had dominated business computing. In 1984, one of the most famous television commercials of all time ran during the Super Bowl. It featured an athletic blonde woman being chased by guards through a meeting hall full of human thralls in a 1984-inspired dystopia. The woman threw a sledgehammer into the televised face of a Big Brother character, freeing the people from the enslavement of the state. The commercial, run by Apple Computers to announce the launch of the Macintosh computer, was a not-so-subtle attack on IBM. It also suggested that computers equaled freedom.

What both Apple and IBM failed to realize was that the battle would not be over hardware but over software. Hardware is important, but what turned the computer into such a powerful tool was the software. Applications such as word processing, accounting programs and databases turned microcomputers into actual tools useful for business, education, research and home use. Although programming computers can be traced back to Charles Babbage, such programs (also known as machine code or low-level programming language) were essentially written in mathematical terms using the specific commands the machine used. It was the invention of high-level languages such as COBOL, Fortran, BASIC and C++ that moved computing out of the laboratory because they used common language terms that were easier to understand. The most famous of the programing pioneers was Grace Hopper (1906–92), who began working on computers at the dawn of the digital age.

 VIGNETTE

Grace Hopper, the Programmer's Programmer

Grace (née Murray) Hopper was a mechanically inclined child and an exceptional student (Figure 11.5). She earned a Ph.D. in mathematics from Yale in 1934 at a time when

social norms dictated that women rarely went to university; if they did, then they did not do science, mathematics or engineering. In 1940, she tried to enlist in the US Navy, but was rejected because of her age. She was accepted into the Naval Reserve and was assigned to

Figure 11.5 Grace Hopper and UNIVAC in a 1961 publicity photograph.

work for the Bureau of Ships Computation Project based at Harvard University, where she worked on the Mark I computer. This was the start of a career in computing. Following the war, Hopper joined the Eckert-Mauchly Computer Corporation and worked on the UNIVAC project.

Much of Hopper's work was on compilers, which are programs that translate written commands into a code that the computer can run as an executable program. She recognized that programming would be much easier if it used common language and employed the power of the computer itself to translate the programming language into executable form. This would also have the advantage of making programming less dependent on any particular computer, since the compiler would do all the work of translating the programming language into the appropriate machine code. Her department created MATH-MATIC and FLOW-MATIC, two early programming languages based on Hopper's ideas. In 1959, Hopper participated in the creation of the programming language COBOL (COmmon Business-Oriented Language). Because the US Department of Defense chose

COBOL as the main programming language for the military, it quickly became an industry standard.

Hopper went on to advocate for distributed computing, presaging the development of the internet. During this time, she remained in the Naval Reserve until she retired in 1966, but she was brought back to active duty. When she retired again in 1986 she held the rank of rear admiral. She was a tireless advocate for science and technology education, and was famous for handing out pieces of wire 30 cm (11.8 in) long to represent the distance light traveled in a nanosecond as part of her lectures on computers. Among her many awards and honors, she was given the National Medal of Technology and posthumously awarded the Presidential Medal of Freedom.

In an age and profession dominated by men, Hopper faced and overcame many challenges that had nothing to do with her work but she rarely commented on that part of her experience. She did say: "Humans are allergic to change. They love to say, 'We've always done it this way.' I try to fight that. That's why I have a clock on my wall that runs counter-clockwise" (Schieber 1987).

At the forefront of the software market for personal computers was Microsoft. Founded in 1975 by Bill Gates (b. 1955) and Paul Allen (b. 1953), it focused on software rather than hardware. Microsoft licensed the MS-DOS operating system to IBM in 1981 when IBM entered the personal computer market, but IBM, believing that the primary financial return was selling computer hardware, agreed to let Microsoft retain the rights to the operating system. Microsoft went on to be the leading supplier of operating systems and business software in the world. In effect, it sold people a program that managed the operation of their computer and then sold them programs such as word processors and spreadsheets that sat on top of the operating system. Together they made it possible for the computer to do calculations, store information and operate devices like printers, scanners and anything that could be connected to a computer. The computer could also run programming languages that allowed users to create their own programs.

Apple and Microsoft developed a complex relationship, first as partners and then bitter rivals, and then as partners again. Apple's computers benefited from the company's drive to make computers more user-friendly. A major development for personal computers was the introduction of the graphical user interface (GUI) that made it possible to put menus, pointing devices such as the mouse and cursor systems and "buttons" on the monitor. These ideas were first worked out by the research team led by Douglas Engelbart (1925–2013) at the Stanford Research Institute during the late 1960s and were further developed at Xerox PARC (Palo Alto Research Center). Xerox did not really know what to do with the innovations, so let Apple use the ideas, which Apple eagerly incorporated into their devices. When Microsoft introduced the Windows operating system, it used a GUI interface that Jobs accused Gates of stealing from Apple. Gates was reputed to have answered: "Well, Steve, I think there's more than one way of looking at it. I think it's more like we both had this rich neighbor named Xerox and I broke into his house to steal the TV set and found out that you had already stolen it" (Isaacson 2011: 178).

⚙ The Internet

The other great innovation that changed the place of computers was the internet. Many people contributed to both the hardware and the software necessary to link computers around the world. One of the pioneers was J. C. R. Licklider (1915–90) who presented the idea of a computer network in his 1960 paper "Man–Computer Symbiosis." When Licklider worked for the US Department of Defense's Advanced Research Projects Agency, in 1962 he linked three computer terminals together, connecting the System Development Corporation in Santa Monica, California with a terminal at the University of California, Berkeley campus and one at the Massachusetts Institute of Technology. The first text messages, the earliest version of emails, were sent in 1965. In 1969 ARPANET was started, primarily linking computers at universities and major research centers, and by 1981 it had more than 200 nods or hosts. This allowed computer computational time to be shared, but also was used to send files and communication. The evolution of the internet was in some senses the linking of many networks. This was made possible by the adoption of a common language of communication, primarily TCP/IP (Transmission Control Protocol/Internet Protocol). In a technical sense, the internet was born on January 1, 1983 when TCP/IP became the only protocol allowed on ARPANET and that in turn made it possible to link any network that used the common protocols to the system.

In 1992, the US Congress allowed commercial activity on the network. This was highly controversial, as the original users had seen it as a scientific and academic tool and objected to commercialization, but as more and more companies joined the network, it became clear that the net was not going to be reserved for specialists. Portal companies, often already in the telecommunications business, offered access to the network to commercial and individual users, usually for a monthly fee. The original connections for individuals were over telephone lines, but as the demands for access rose and the amount of information being sent climbed, higher-capacity connections using coaxial cable and fiber optics were introduced. The availability of email as a cheap method of communicating was very attractive, and with the addition of graphics and browsers the range of material that could be sent to people over the internet turned it into an enormous resource for everything from business management to cooking lessons.

The internet was transformed by three linked "languages": HTML (hypertext markup language), CSS (cascading style sheets) and JavaScript. Together, they allow information to be transmitted, stored and displayed by any device that can run the languages. To get the information, the Hypertext Transfer Protocol (HTTP) governs the process of requests and responses. For example, typing a web address into a browser sends a request to a server, and in turn the server responds by providing the HTML files and the website appears on the screen. What makes the system so complex is that, unlike a telephone where a direct connection is made, the sender and the receiver have to know not only where each other is on the network but how to send information via many possible paths. That information is broken up into blocks or packets, and not all the packets need to travel the same path from the source to the final destination over the network, but they need to be assembled in their proper order to work. This requires complex mathematics, but, in a kind of self-reinforcing technological development, we have powerful computers to calculate and manage the flow of information.

With such vast amounts of information available, it became difficult to find things, so one of the most significant developments on the internet was the appearance of search engines. The first was Archie, created at McGill University, Montreal in 1990, but there were soon dozens of general and specialized search engines. For general searches, by about 2000 Google, founded in 1996 by Larry Page (b. 1973) and Sergey Brin (b. 1973), became the most popular search site on the internet. A large part of the success of Google was based on the ability of the company to generate profits from their engine. This allowed the company to increase its utility and add new components such as email, text translation and calendar functions through their online portal.

The online world has also changed the definition of privacy. Unlike physical letters and old-style telephones on exchanges, both the content and the "meta-data" (data about things like the addresses of senders and recipients, aggregated information about sites visited, purchases, or even word use) of online communications are available for capture and analysis. Statistics from search engines, for example, have been used to track the spread of the flu as people in affected areas use the internet to look for information on symptoms and treatment. What you post on social media sites is being used to determine what advertising you get. Governments are increasingly using mass information gathering to look for terrorists and criminals, often with little or no oversight. Governments have also moved to outlaw encryption or asked the providers of encryption software to create secret keys or back doors to allow them to inspect encrypted communication. The legal issues are still being worked out.

One of the curious results of the vast oceans of information that have been created by computers and the internet is the decline in common experiences. Some theorists have pointed out that shared interests are replacing proximity as the foundation for communities. Sociologists and others interested in the effect of the internet have pointed out that, despite having almost instant access to a vast collection of information, people tend to look at a small number of websites and seem to be creating and living in information "bubbles." Inside the bubble, only certain kinds of news, political views, music and entertainment are allowed. Even dating in the real world has had a digital makeover as people are pre-screened to have similar political, lifestyle and cultural views. Although conspiracy theories are as old as human politics, in the internet age bizarre ideas can be spread at the click of a button.

The internet, combined with mobile telephone technology, has made the use of social media a major part of modern life. At the most basic level, platforms such as Facebook, YouTube and Twitter have allowed people to connect with friends, family and people with similar interests almost anywhere in the world. Social media have also been associated with social movements such as the Arab Spring, a series of protests, uprisings and civil wars in the Muslim world in the early 2010s. Social media are now a significant part of political discourse, both for campaigning and for governments to get information out to the populace. Yet there is a price for all this. To pay for the systems which are often offered free to users, the information of the users is collected and sold to advertisers, political parties and others interested in tracking people's behavior, buying patterns or social interests. In 2018, a series of scandals revealed the secret use of Facebook account information by Donald Trump's presidential election campaign and by the pro-leaving side of the Brexit (Britain exiting the European Union) Referendum.

To make matters even more confusing, the internet has made it possible to reveal information in a way that is very hard for governments or businesses to control. The organization Wikileaks operates the most famous website that discloses secret information from governments, intelligence services and large corporations. For many people, this has made the world more open and knowable, but it has also made it easier to spread false information. In the vast ocean of information that the internet has made available, the distinction between imaginary ideas about the world and real information has been eroded for many people. The internet has made it much easier for people to create an information bubble for themselves, choosing to only look at information that supports their political or social ideas, but they can also have the bubble created for them by algorithms that filter what users see, based on past choices such as internet searches or movie selections.

This has been accompanied by the rise of "fake news," which is a form of propaganda dressed to look like traditional news. The internet and social media have made it easy to spread false news that has all the appearance of reliable sources, with beautifully designed websites and high-quality audio and video. Fake news has been used as a tool to influence public opinion in over twenty countries. The concept of fake news has also been used as a way to discredit actual news, as politicians such as President Donald Trump and President Vladimir Putin accuse opponents of spreading false stories that are actually true.

⚙ Energy: Networks of Power and Dependence

The foundation of the modern world, especially in the digital age, is built on the exploitation of energy. For thousands of years, the main source of fuel used by humans was wood, either in natural form or as charcoal. Other organic fuels such as peat, dried dung, pitch and coal were also used, depending on local availability. In the nineteenth century, coal replaced wood in the industrialized world as the primary energy source for everything from home heating to heavy industry such as steel making. In the early twentieth century, electricity and petroleum products began to replace coal as the pre-ferred form of energy for end users, although petroleum and coal were both used to generate electricity in places where hydroelectricity was unavailable.

Hydroelectricity offered a very low-cost form of power since the motive force was simply fast-flowing water turning turbines. Governments and businesses began to see rivers as vastly under-utilized sources of power and as a way to make power and money that had a virtually unlimited lifespan. The first of the hydro mega-projects was the Hoover Dam on the Colorado River in the United States. Completed in 1936, it was to provide both power and flood control. The golden age of hydro projects began in 1957 with the completion of the Samarskaya Dam on the Volga River, and a new dam was being completed somewhere in the world almost every year until construction slowed in the 1980s. The decline was largely because the rivers that could be easily used for hydro in Europe and North America were all in use.

Politics and power came together with the building of the Aswan High Dam on the Nile in Egypt. In 1952, the Egyptian Revolution overthrew the Egyptian monarchy and brought Gamal Abdel Nasser to power. The new government wanted to industrialize Egypt and to do that they needed a reliable and inexpensive source of electricity. There was a small dam on the Nile, but it was old and at its capacity, so a new, much larger dam would provide power and control flooding. Initially Britain and the United States offered to fund and help build the dam, but tied their support to other political and mil-itary demands on the Egyptian government. Nasser accepted an offer from the Soviet Union and construction began in 1960 and finished in 1970. The dam represented a number of conflicting interests, such as the struggle between the USSR and the West during the Cold War, but the Aswan dam was also an attempt to use technology on a large scale to transform a society. This set agrarian interests against industrial interests and led people to ask what the price of modernity would be in terms of loss of culture and tradition. Did becoming an industrial state mean becoming either a capitalist or a communist state?

The Aswan High Dam was also one of the first dams that generated serious concerns about environmental damage, first because it flooded a valley full of ancient Egyptian artifacts that had to be moved, but also because the dam basin filled with silt. The silt has to be dredged and dumped to keep the dam working. In the long term, the Nile delta is eroding because the silt that renewed it no longer reaches the delta, and Egypt is using artificial fertilizers for farming to replace the renewal of the soil that the annual flood used to provide. There is a certain historical irony that the feature of the Nile that made ancient Egypt such a powerful empire was seen as an impediment to making Egypt a modern industrial nation.

Starting in 2000, there has been an explosion of hydroelectric projects in one country. China has built twenty-one major dams, including the highly controversial Three Gorges Dam on the Yangtze River. This massive electrification strategy is largely about supplying power to industry and a growing domestic market, but it is also part of an effort to get China off coal, which in combination with the rising use of automobiles has produced some of the world's worst air pollution.

Nuclear Power: Promises and Problems

A new power technology that grew out of the research on nuclear weapons during the Second World War was nuclear energy. Although scientists had known since 1896 that radioactive materials produce energy, it took until 1942 to find out if that energy could be controlled. In that year, Enrico Fermi and his team at the University of Chicago built the first atomic pile as part of the Manhattan Project. The creation of the pile was important because it confirmed that nuclear chain reactions were possible and allowed the scientists to calculate the amount of energy that could be generated by nuclear decay. This was the first step toward building atomic weapons, but it also demonstrated that atomic energy could be used to produce power, since the heat produced by the atomic pile could be used to do work in the same way that the heat produced by burning coal or oil could do work.

The first nuclear power plant to generate electricity began operating in 1954 in Obninsk, Russia. It was a small experimental reactor that used graphite as a moderator (controlling the rate of neutron interactions with nuclei) and water cooling. It made steam to spin a turbine and generate electricity. In Britain, the Calder Hall test reactor opened in 1956, and the Shippingport Atomic Power Station, in Shippingport, Pennsylvania, the first nuclear plant designed just as a power generator, began operation in 1957. Because nuclear power offered the ability to generate power for long periods without refueling, it was ideal for naval ships, and in 1954 the US Navy launched the nuclear-powered submarine *USS Nautilus*. In 1958, she traveled under the sea ice to the North Pole.

Nuclear power was seen by many as the energy source of the future. Plants could be built anywhere and they produced electricity with no air pollution, very low greenhouse gas emissions and less environmental impact than coal power. In principle, the technology of nuclear power was simple. The heat generated by the decay of uranium or other nuclear material was used to heat water to make steam. The steam turned a turbine and generated electricity. A cooling system (usually water cooling) converted the steam back to water and the water was recycled into the reactor (Figure 11.6).

In practice, a number of factors limited the utility of nuclear power. First, nuclear plants are very complex machines with enormous investment costs and require technical support staff with high levels of training. Second, nuclear power works best when generating a constant flow of electricity, and cannot easily be turned on and off as demand changes over the course of the day. Third, the plants create nuclear waste that is dangerous and long-lasting.

These three problems have technical solutions, but the biggest constraint for nuclear power is not the technical issues about building and running nuclear plants but the fear of catastrophic failure. These concerns were reinforced by a number of accidents.

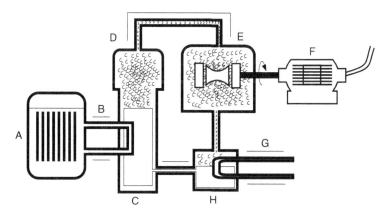

Figure 11.6 Nuclear power generator. A The reactor vessel heats up fluid that passes through B and turns water into steam at C. Steam passes through D and turns a turbine E. The turbine powers a generator F that produces electricity. Cooling water G converts the steam back to water at H, completing the cycle.

A number of early accidents such as the 1957 Windscale fire in Cumberland, United Kingdom and the 1969 Lucens reactor meltdown at Vaud, Switzerland received some publicity, but the first large-scale and commercial accident that brought the issue into public discourse was the Three Mile Island accident in 1979. Three Mile Island Nuclear Generating Station (TMI-2) in Pennsylvania suffered a mechanical failure in its cooling system, and a combination of poor monitoring equipment and operator errors led the operators to believe that too much coolant was in the system when coolant had actually leaked out. This led to the release of radioactive material and triggered a voluntary evacuation notice that resulted in more than 140,000 people leaving the area.

In 1986, the worst nuclear plant disaster happened at the Chernobyl Nuclear Power Plant near Pripyat, Ukraine. A combination of design flaws (in particular a back-up cooling system that took too long to start) and operator error led to the core overheating, followed by a steam explosion and fire. Radioactive material was released into the atmosphere and precipitated across Europe. Thirty people died as a direct result of the explosion and fighting the fire, and an estimated 9,000 will die prematurely from radiation exposure. A large area around the plant has been abandoned, and the long-term effect of the radiation release is still being monitored.

The most recent nuclear accident was at the Fukushima Daiichi Nuclear Power Plant in Japan in 2011. Following an earthquake, a tsunami struck and flooded the plant. Although emergency procedures shut down the plant because of the earthquake, the flood knocked out the emergency generators keeping the core cool. This led to core meltdowns in three reactors and a hydrogen explosion. No one died from the accident, but about 1,600 deaths have been attributed to the evacuation, and radioactive material is currently still seeping into the ground.

Frightening as these events are, a number of scientists and engineers have pointed out that nuclear power, even with these accidents, kills far fewer people than any other non-passive energy source. Coal is the deadliest energy source, killing between 2.8 and 32.7 people per billion kilowatt hours (depending on whether you are counting only direct fatalities from industrial activity or including deaths from pollution and disease). There are only 0.2 to 1.2 deaths per billion kilowatt hours from nuclear power (see McKenna 2011: 23). Thus, nuclear energy poses a complicated problem as a technology. In a world hungry for energy but concerned about pollution and radiation exposure, nuclear power offers both benefits and challenges. Nuclear plants are among the most complex machines that have ever been built, which means that they are the most complex machines to maintain. The three worst accidents have happened in countries with strong technological infrastructures, so we could reasonably ask what might happen to a nuclear plant in a country in crisis or that lacks the ability to maintain the technological infrastructure?

Anthropocene: Understanding Technology on a Global Scale

Industrialization has been made possible on a global scale because we can generate energy from carbon-based fuels. In turn, the technology of the industrial era gave humans the ability to control and exploit the environmental on an equally global scale. The intrusion of human activity into every ecological niche was called the "Anthropocene" by Paul Crutzen and others as a way to signify the effect humans were having on the environment.

The Anthropocene covers a number of environmental changes brought about by human activity, but the most significant is climate change due to global warming. The pollution associated with carbon fuels and other industrial sources of greenhouse gases has been identified as the primary driver of a sudden and potentially catastrophic trigger for climate change. The idea that heat could be trapped by Earth's atmosphere was first proposed by French physicist and mathematician Joseph Fourier (1768–1830) in 1824. His idea was tested by a number of scientists, and in 1896 Svante Arrhenius (1859–1927) made the first prediction about a warming Earth. The process was called the "greenhouse" effect by Nils Gustaf Ekholm (1848–1923) in 1901 and it was originally linked to geo-climate studies that were attempting to explain the fluctuation of warm periods and ice ages over millions of years. By the 1960s there was growing concern amongst a small group of scientists that the vast amounts of carbon dioxide (CO_2) coming from modern industrial processes and transportation were replicating the conditions that were seen in the ancient warm periods. Measured from the beginning of the Industrial Revolution and taking into account natural sources for heat accumulation such as increased solar radiation

and volcanic events, the inevitable conclusion was that human activity, particularly the dumping of gigatonnes of CO_2 into the atmosphere, was heating up the planet faster than any natural process.

Higher temperatures mean changes to the physical behavior of all the climate systems, including ocean currents and salinity, the disappearance of glaciers and ice caps, rising sea levels, changes to the air circulation patterns, and an increase in the power of storms as more energy is available for weather systems.

The problem is clear, but the solution is not. Technology associated with industrialization and improving economies carried with it an unexpected danger that functions on a global scale, while the ability to control those technologies is at best national. Further, controlling technology seems to be at odds with the desire for economic development. Getting all the nations and the industrial interests to agree to any program of actions is an enormous political problem. This has been made more difficult by a small but influential group of opponents to the idea of climate change, despite the overwhelming evidence. The situation has many people thinking about Rapa Nui and Çatalhöyük, where the people of those civilizations perfected their technological ability to exploit the local environment, but by doing so stripped the environment of the very things that allowed them to develop their civilization and its technologies in the first place.

If we assume that we are wise enough to recognize the reality of the problem, the question then becomes: How do we get out of the trap? There would seem to be only two paths out: Get rid of most technologies, or use technology to solve the problem.

We could abandon modern technology and return to pre-Industrial Revolution ways of living, but the consequences of that would be enormous, since the vast global networks of transportation, manufacturing and agriculture would cease to exist. Populations would, by necessity, fall to the point where pre-industrial food production methods could sustain the population. There would have to be international agreement and enforcement to ensure adherence to such a plan and that seems unlikely. The people most skeptical about the possibility of international action foresee a global economic and environmental collapse that will enforce a return to pre-industrial life.

At the extreme end of technological solutions to climate change is geo-engineering. This would require solutions on a planetary scale, such as reflective shields in space, mass sequestering of industrial greenhouse gases, and the use of genetically modified plants to capture carbon from the atmosphere. In combination with new forms of low-carbon energy production, geo-engineering offers the possibility of addressing climate change with little change to the lifestyle of the industrial world. The costs of such projects could be astronomical, but the prospect of maintaining industrial activity might make geo-engineering attractive despite the cost. We would, however, be relying on unproven technology with unknown consequences and we would still have to get global agreement. Even proponents of geo-engineering point out that there are limits to natural resources, so we would still have to deal with problems of population growth and ultimately how to get resources from asteroids or other planets.

What seems more likely is a combination of new technologies, some modification of lifestyle particularly in the industrialized world, and adaptation. A concerted effort to limit population growth, for example, would reduce the demand for resources, while improving energy production from non-carbon sources would reduce carbon production. Globally, we would still face big problems, such as the loss of coastal cities, the shifting of agriculture and environmental destruction, but we would avert a complete collapse of civilization. Over time, we would learn to balance human activity with the environment at some equilibrium level.

The prospects of the Anthropocene era seem dire, but we have a number of things available to us that the ancient civilizations did not. First, we have the technology to recognize the problem. Second, we have the lessons of others who suffered from environmental problems to help us understand the technological trap. Third, and perhaps most important, we already have a global community. Simply the ability to recognize that our problems are not local, and that what happens on the other side of the globe can affect us, gives us a major advantage over the people of previous ages. For example, we already have the technology to deal with many of the problems of energy production, but we may lack the political desire to pay the cost of using that technology. We can see both the potential and the problems of dealing with climate change by looking at the Paris Climate Accord. In 2016, 174 countries and the European Union signed on to this. The signatories represent all the major industrial nations and most of the nations of the world, suggesting that global action is possible. On the negative side, most of the signatories have so far failed to meet their own reduction targets, and the United States, the second largest producer of greenhouse gases after China, has announced its intention to withdraw from the Accord.

✿ Our Battery-Driven World

Another delivery system for power is the battery (Table 11.2). The introduction of so many new electronic devices increased the demand for batteries, particularly for portable

Table 11.2 Battery history.

Date	Inventor	Innovation	Comment
1836	John Frederic Daniell	Dual electrolytes	Eliminated hydrogen generation
1844	William Robert Grove	Platinum cathode and dual electrolytes	Greater voltage
1859	Gaston Planté	Lead-acid battery	Rechargeable
c. 1860	Callaud	Gravity cell using zinc sulphate and copper sulphate	State easily determined, renewable, high charge
1881	Camille Alphonse Faure	Lead grid and lead oxide paste	Improved performance, easier mass production
1866	Georges Leclanché	Zinc anode, manganese dioxide cathode wrapped and dipped in ammonium chloride	Ease of use, good charge
1887	Carl Gassner	Dry cell using paste rather than liquid electrolyte	Portable, no acid spills
1899	Waldmar Jungner	Nickel-cadmium, first alkaline (not acid) electrolyte	Portable, rechargeable, long lasting
1899	Waldmar Jungner Thomas Edison	Nickel-iron	Powerful, used for early electric car
1959	Lewis Urry	Zinc-carbon	Low-cost alkaline battery
c. 1970	US Navy	Nickel-hydrogen	High-power, long charge life but expensive
c. 1970	G. N. Lewis	Lithium	Light, long charge, expensive
c. 1980	John B. Goodenough	Lithium ion	Rechargeable, long charge
1989	Various companies	Nickel metal-hybrid	Lower cost than nickel-hydrogen
1996	Various companies	Lithium ion polymer	High charge, can be almost any shape
c. 2007	CSIRO UltraBattery	Lead-acid ultracapacitor hybrid	Large storage and automotive
c. 2009	DARPA	Silicon-air	Still experimental

devices such as transistor radios and portable computers. There are two classes of batteries, known as primary cells which are non-rechargeable and secondary cells which can be recharged. Large batteries, such as those used in automobiles, could supply more than enough power for electronics but were too big, while the traditional dry cell battery that was created at the end of the nineteenth century and used for simple electrical devices like flashlights and toys did not carry sufficient charge and had to be replaced frequently. The solution was to use more exotic materials such as lithium for batteries that could be recharged. The new secondary cells have freed many devices from direct connection to electrical power supplies, but that freedom comes at a cost. Batteries are expensive and must still be connected to a power supply to be recharged, and there can be problems with toxic waste. About 180,000 tons of batteries are discarded in the United States annually.

⚙ The Integration of Everything

Batteries have partially liberated electronic devices such as smart phones, tablets and laptop computers from the power grid, and promise the possibility of transforming the way we use power in our homes, businesses and transportation. We still need to connect to the grid to recharge batteries, but renewable energy generation and localized electrical production combined with new generations of storage batteries will change the power grid itself and offer a level of self-sufficiency.

It may seem paradoxical, but the new devices that are starting to disrupt the established networks of power transmission and communication that link a huge portion of the world's population together have been made possible by the creation of massive new systems of integrated devices. The integration of telephone systems, mobile phones, cable television and the internet has produced the most massive network ever created. Just the ability to integrate systems based on copper wires with coaxial cable, fiber optics, radio frequency and microwave transmitters to satellites represented the invention of tens of thousands of devices and software programs. In addition to the technical problems of integration, we must remember the human part of the creation that required the work of inventors; the adoption and production of tools, machines and devices by competing companies; and the agreement of multiple governments and business leaders about certain standards such as what frequencies could be used by mobile phones, or what URL addresses would be recognized.

One of the results of this integration has been that the distinction between devices starts to disappear when you can listen to a radio program, watch television, turn on the kitchen lights and call a friend on the same device. That device might look like a television, a cell phone or a laptop computer, but these are all simply providing access points to the communications network.

While the internet is generally conceived of as a system to link people, and in fact billions of people are connected to the global communications system, its greatest impact is the ability to connect billions of *devices*. Everything from radio telescopes to electronic door locks is connected. Sometimes called the "internet of things," we are connecting more and more non-electronic things to the network, such as running shoes, footballs, bathtubs and refrigerators. Part of the reason we can do this is that computer languages such as HTML (hypertext markup language) and Java have been created to allow the integration of network functions (such as how you send and receive information) with applications (anything that a specific device does) across multiple platforms. Thus, Java is used by desktop computers, Blu-ray players, car navigation systems and 3 billion mobile telephones.

The creation of the computer and the internet has transformed communication and information distribution. It is certainly the case that the computer revolution has created not just new jobs, but whole new types of jobs. Herein lies one of the potential traps of the modern world. If so much of the world is controlled by computers and the networks they operate, do we lose something in exchange for the benefits? The digital age has brought much of the world closer together, so that we can see and hear, explore and buy things from faraway places. It has also meant that we are even more dependent on systems that are so technologically complex that no individual can understand all of it, and we need highly trained people to keep everything going. A neo-Luddite might suggest that the internet is the modern-day equivalent of the flour mill in Roman Barbegal. If there is a disruption in the larger community, the skills to keep the computers going might disappear. Unlike Barbegal, the technology of computers cannot be figured out by observation of the devices themselves, so recovering from a second dark age might take considerable time.

On the other hand, we might be on the verge of creating a vast, global community that unites us into a more peaceful and prosperous world. One of the astonishing things about the internet is that it contains vast resources about the internet. All of the intellectual component parts are available, such as the various languages (many of which are open source or free to use), instructions on how to write code, and communities of people willing to share their knowledge. Never in human history have so many people had access to so much information, and that offers us both opportunities and problems to solve.

1 How did the telegraph and telephone lead to the centralization of economic and governmental power?
2 What was the "Green Revolution?"
3 Why might we divide the history of technology into "pre-computer" history and "post-computer" history?

NOTES

1. The name was suggested by John R. Pierce and combined the words "transconductance" and "varistor," the latter an electronic device that had variable electrical resistance depending on the voltage passing through it.
2. The earliest reference to portable field radios being called "walkie-talkies" appears in the *New York Times* in August 1938.
3. As a comparison, it cost about $166 (2010) to ship one ton of bulk cargo from New York to England in 1800.
4. In comparison, the Boeing 707 carried between 140 and 219 passengers or 81 tonnes (90 tons) of cargo.
5. The term "Moore's law" was coined in 1970 by Carver Mead.

FURTHER READING

The rate of invention increased so fast during the twentieth century that thousands of objects have become the subject of investigation. In the broadest terms, however, the introduction of digital technology has changed the relationship between society and objects. In communications, R. W. Burns' *Communications : An International History of the Formative Years* in *IEE History of Technology* 32 (2004) is widely available as an ebook and is a sweeping outline of the past to the television age. Looking at one of the most important inventions, Christopher Beauchamp raises an interesting historiographical question about how to understand the process of invention in "Who Invented the Telephone? Lawyers, Patents, and the Judgments of History" (2010). Building on the introduction of electronics pioneering in communications is Martin Campbell-Kelly and William Aspray's *Computer: A History of the Information Machine* (2014) which looks at the broad history of the computer. As a source of online information, visit the Computer History Museum at www.computerhistory.org. The intersection of computers and people is examined by Lisa Nocks in *The Robot: The Life Story of a Technology* (2007).

12 Conclusion: Technological Challenges

1895	Konstantin Tsiolkovsky suggests a space tower, later called a space elevator
1957	Launch of Sputnik I and II
1959	Richard Feynman introduces idea of nanotechnology
1962	Telstar I first communications satellite
1969	Apollo Moon landing
1971	US Supreme Court decision opens the way for genetic patents
1984	Stereolithography or 3-D printing
1985	Beginning of carbon nanoscience

Technology has given humans the power to control the world around us in ways our ancestors never dreamed possible. In the twenty-first century, new forms of technology have allowed us to make very small nanobots and plan Mars colonies. It sometimes seems like we are living in a science fiction world. Particularly for those people fortunate enough to live in the industrial world, we are surrounded and protected by vast networks that supply food, power, transportation and communications. Yet technology can also be a trap. The benefits of technology are not equally distributed and our dependence on power, particularly non-renewable power, makes all of us vulnerable to the failure of energy supply. Pollution, overpopulation, limited resources, climate change and political turmoil all have technological roots. Solving these problems will require the ability to understand what technology offers us.

Clarke's Third Law: Any sufficiently advanced technology is indistinguishable from magic.   (Clarke 1973)

At a profound level, we are our technology. Humans live and prosper because we take part in networks of physical and social technologies. Whether we are using stone tools or complex computers, we interact with the world in certain ways because of the devices we have. Yet it is not the artifacts that tell the whole story of technology; even more than the devices themselves, we are technological beings because we accumulate knowledge and share it with others. We organize ourselves to take on tasks that are beyond the scope of an individual, and share in the rewards that come from combining our skills and abilities. There are dangers associated with our technological life, such as pollution, the devastation of industrialized war, the destruction of existing social structures when new technologies are adopted, and the ever-present potential to exhaust the natural resources that support civilization. Yet the history of technology, even with the rise and decline of various civilizations and cultures, has tended toward more complex and larger networks of human cooperation and the production of larger and more complex material goods.

Importing Technology

One of the problems with a global market for technology has been the effect of imported technology. This is a form of technological determinism where the people involved believe that by importing technology they can change their society in a specific way. There are benefits to such transfers, such as avoiding the cost in time and money of developing technology that already exists elsewhere, but there are also risks. Imported technology can fail because of a lack of physical, intellectual or cultural infrastructure or have unexpected consequences.

Some notable attempts to import technology include various European leaders trying to lure craftspeople such as the Murano glassmakers of Venice or the lacemakers of Flanders to their regions as a way of improving the economy and gaining technically skilled people. In 1724, Emperor Peter the Great tried to jump-start Russian science and industry by creating the St. Petersburg Academy of Sciences and bringing in many foreign scholars. During the Meiji period (1868–1912) in Japan, there was a concerted effort to replicate European or American industries. In more recent times, we could look at the Aswan High Dam and the spread of nuclear power plants to less industrialized countries as examples of technology that has been imported to regions that did not have the bureaucratic or material infrastructure to create the technology themselves.

The effort to import technology has had a variety of outcomes. Japan's rise to regional and later global power can be traced to the effort to import technology, but it was accompanied by periods of serious social upheaval. It took over a century for the Academy of Sciences to have much impact on science in Russia and the expected practical applications never materialized. The Aswan High Dam really did help Egyptian efforts to industrialize, but the effect was uneven, with some people benefiting far more than others. Nuclear power plants in many less industrialized countries have often needed the ongoing support of foreign governments and companies to keep operating.

✿ Bhopal: Necessity and Disaster

In 1969, Union Carbide India Limited (UCIL) opened a pesticide plant in Bhopal, India. The company was a joint venture of the Indian government and the US-based Union Carbide Corporation and was part of India's effort to increase agricultural production known as the Green Revolution. The changes to agriculture were partly based on the development by Indian scientists and agronomists of new crops, but changes also depended on industrial farming methods imported from the West, including mechanization and the use of fertilizers and pesticides. In 1984, more than 500,000 people were exposed to methyl isocyanate and other toxic substances from a gas leak at the UCIL plant. The number of deaths is highly disputed, running from a low of 2,300 to a high of 16,000. Tens of thousands of people suffered injuries ranging from skin rashes to permanent lung damage.

UCIL had a history of leaks dating back to 1976 and workers suffered a number of injuries and one death from chemical exposure. When the disaster occurred, there were a number of safety systems in place, but they were turned off, were not working or were insufficient to deal with the volume of chemicals being used. On December 2, a tank holding the methyl isocyanate overheated and vented about 40 tonnes (44 tons) of chemicals into the air over the course of two hours. A siren meant to alert the public to gas was turned off because the workers believed they could deal with the leak and did not want to panic people. When police called the plant about gas complaints, they were told that there was no problem. It was only after the gas cloud had already blown into the city that anyone was warned of the leak, by which time it was too late.

There has been a great deal of analysis of the Bhopal disaster and blame for it has been directed at everyone involved. The Indian government failed to oversee the plant, the parent company seemed unaware of or unconcerned about the safety problems, there were too few workers and they were undertrained, the plant design was poor, the location of the plant was far too close to a populated area, the managers were more concerned about production than safety, and no one seemed to be aware of the life expectancy and maintenance cycle of such a plant. These issues all relate to who was responsible for the disaster, but we can ask a more basic technological question: What real-world problem was the plant supposed to solve?

The answer to that question reveals the complexity of technology transfer. India needed to maximize its food production, and just like Germany at the beginning of the twentieth century, it turned to technological solutions including mechanization, improved transportation systems, fertilizers and pesticides. This effort has worked. Between 1950 and 2010, crop production in India rose from about 50 million tons annually to around 220 million tons (see Cagliarini and Rush 2011). Unlike Germany and other Western countries where the technology grew from small-scale operations to large-scale use over several decades, India imported the apex technology and imposed it on the existing agricultural system. Although greed and incompetence played a role in the Bhopal disaster, the lack of a history of dealing with the technology also played a role. European and American chemical producers all had accidents, but the accidents mostly happened during the developmental stages of the industries and were thus smaller. The reaction to the problems resulted in systems and regulations to limit disasters and significant legal penalties for infractions.

Another difference between India and the West is that most of the regions that industrialized in the nineteenth century had seen a decline in population growth and even some areas of falling population by the start of the twenty-first century. India, although it possesses a significant industrial sector and has one of the biggest economies in the world, has continued to experience population growth and constant pressure to keep food production up. That means that simply outlawing chemical pesticides to avoid another Bhopal is unlikely. Better production methods and safer products can mitigate the dangers, but population pressure will keep India and other countries looking for technological solutions.

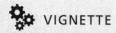 VIGNETTE

Collapse: When Societies Lose Their Technological Systems

Most of the story of the development of technology is about the ability of humans to use their creativity, cooperative abilities and knowledge of materials to solve real-world problems. The successful use of technology has enabled population growth, increases in the standard of living and the spread of education to people around the world. Yet it can also be a trap. History can teach us lessons about the problem of technological dependence, especially when technology becomes critical to society while at the same time internal or external events degrade or destroy a society's ability to maintain the underlying systems on which the technology depends.

The most common external reason for collapse has been unexpected changes in climate. Drought hit the Harappa civilization in the Indus valley around 1900 BCE and the Mayans around 900 CE. Both groups were dependent on local agriculture for food and both cultures attempted to deal with the droughts by building irrigation systems. The technological solution helped for a little while, but as the droughts stretched on and famine and social strife reduced populations and wrecked social organization, it became impossible to undertake any systematic response to the problems.

In the cases of the Anasazi people in the American Southwest and the people of Çatalhöyük in Turkey, the creation of large settlements contributed to the collapse. The ability of the people to exploit local resources was so good that they deforested the region and ruined the water resources,

and that in turn led to the collapse of social order and violence. A similar story can be told about the people of Rapa Nui, who had the technological skills to build massive stone statues, but by cutting down all the trees on their island (rats also ate the seeds so new trees couldn't grow) the people could no longer build boats or shelters and began to burn grass for fires. To create the tools (especially boats and shelter) needed to support their way of life actually wiped out the source of those tools.

The fall of Rome, referring to the collapse of Roman civilization in western Europe, had a number of contributing factors, including climate change, corruption of government, and war, both civil war and invasions. Yet another part of the collapse was technological. The Roman Empire depended on the movement of food, people and materials over long distances. To achieve that, information had to be communicated to coordinate and control everything. As soon as the communication system was interrupted, what had been a massively integrated system fragmented into locally controlled regions. These smaller social groups shifted from production for the empire to subsistence production, and within two generations the skills associated with maintaining the integrated empire were forgotten. It took over 800 years to regain all the technical skills that were lost with the fall of Rome.

Technological decline can also be the direct result of the destruction of places and organizations that traditionally were the sources of innovation. When the Mongols rode out of central Asia, they destroyed Baghdad and the House of Wisdom in 1258.

Many historians see this as a turning point in the Islamic Golden Age, noted for its high levels of artistic and engineering creativity. Although the Islamic world did not undergo the level of collapse that affected the Roman Empire, there followed a slow decline in interest in the natural world and new technology. Eventually, the Ottoman Empire fell behind its European neighbors in terms of innovation, losing territory and influence throughout the nineteenth century and disappearing in 1922.

The vulnerability of complex systems can be seen in the Great Depression that started in 1929. The specific start to the depression came with the collapse of the stock market, but the depression was inevitable. The First World War had damaged the economy of Europe and the only way to recover seemed to be through industry. The economies of the industrial countries were based on mass production, but everyone was selling and no one was buying. To protect domestic industries from cheap (and often below-cost) goods, many countries imposed high tariffs such as the US Tariff Act of 1930, known as the Smoot-Hawley Tariff. These protectionist policies made the situation even worse. Without markets, workers were laid off and that reduced the market for goods even further. Financial systems were inadequate to control continental and global financial activity, banks went out of business and governments were often paralyzed about

how to address the problems. People could not understand how the failure of a German farmer to get fertilizer could mean that someone working on the assembly line at the Pullman Car Company in Chicago lost their job, but the link was the vast interconnection of commercial interests. When one segment of the system failed, it dragged everything else down as well.

Today, we have a global economy based on the movement of information and physical materials around the world that is several orders of magnitude more complex than existed in 1929. In some ways it is more resilient than the systems of the Mayans or the Romans, and even during the Great Depression. Local problems such as drought are not as devastating as they were in the past because resources can be delivered from other places, but our technological society is still resting on some very slender supports. In particular, our global dependence on energy means that any interruption in the flow of fossil fuels or electricity (which may be generated from fossil fuels) cannot be remedied quickly. Any major failure of our communication systems would throw the world into chaos. And we should remember the Mayans, the people of Rapa Nui and the Harappa civilization. They faced local climate change that wiped out their way of life, while we face a global climate change that touches everyone.

✿ Space as the New Frontier

Rockets and nuclear weapons had been combined to transform warfare in 1957 with the launch of Sputnik I and II. Although the launch of Sputnik I on October 4, 1957 was officially part of the multinational research program of the International Geophysical Year, it was a political and technological statement that the USSR was capable of making intercontinental ballistic missiles. Nuclear missiles eliminated the concept of a front line in global warfare, since no point on the planet was beyond their range. The technology for total extinction of all life on Earth was created and made using the same industrial systems for mass production that produced the Model-T Ford.

The USSR went on to launch the first animal into space, sending Laika the dog on a one-way trip to orbit. On September 13, 1959, the Soviet spacecraft Luna 2 reached the Moon. On April 12, 1961, Yuri Gagarin became the first person in orbit, returning safely. John Glenn was the first American to orbit the Earth on February 20, 1962, but the USSR replied by sending Valentina Tereshkova, the first woman in space, around the Earth forty-eight times in 1963.

The space race was on.

The struggle between the USSR and the USA to be the first to accomplish feats of space flight was partly military and partly political. Prior to Sputnik, the Western powers believed that they were more technologically advanced than the USSR. Spaceflight was big news and seemed to demonstrate that the USSR was a technologically advanced nation and ahead of the West. This proved to be a propaganda coup for the USSR, but it was also a declaration about military power. The same equipment that could launch a person into orbit made ICBMs (Inter-Continental Ballistic Missiles) possible.

On May 25, 1961, President John F. Kennedy announced:

I believe that this nation should commit itself to achieving the goal, before this decade is out, of landing a man on the Moon and returning him safely to the Earth. No single space project in this period will be more impressive to mankind, or more important in the long-range exploration of space; and none will be so difficult or expensive to accomplish. (Kennedy 1961)

The Moon landing program was already being developed by NASA (National Aeronautics and Space Administration) when Kennedy made his pledge, but it became a national priority. Kennedy, assassinated in 1963, would not see the end product of his initiative. After years of development, testing and a deadly fire that killed the crew of Apollo 1, the crew of Apollo 11 reached the Moon on July 20, 1969 and Neil Armstrong became the first person to walk on the Moon. Millions of people around the world watched televised images of the event. When the program ended in 1973, six Moon missions had succeeded and America had reclaimed its technological superiority.

Yet even as the Apollo Program was being developed, there were questions about its utility. There were some scientific questions that the program helped to answer about the Moon and the effects of spaceflight, but the first and only scientist to visit the Moon was the geologist Harrison Schmitt on Apollo 17, the last mission to the Moon. Science was important to NASA, but the race to space was far more about technical prowess. NASA has always promoted itself as an engine of technological development. Starting in 1976, it published an annual report on its technology called *Spinoff*. It lists more than 1,920 products developed by or in conjunction with NASA, including temper foam (better known as memory foam), aircraft de-icing systems, and solar panels. While it is true that NASA is a focus for technological development and represents the largest ever collection of engineers and scientists in a single organization, it is not as clear that it was the best way to gain those technological developments. What NASA is always careful about discussing is its contributions to military technology. The public face of NASA is civilian, but its existence and the core of its work are military, including launch vehicle design for missiles, radar systems, guidance systems and spy satellites.

An important aspect of the space race was missile technology, but in the longer term it is actually the command and control systems made possible by satellite technology that have had a much bigger effect (Table 12.1). Compared to the headline-grabbing manned missions, satellites rarely made the news, but by 1961 there were more than a hundred satellites orbiting the Earth. In 1962, the US launched Telstar I, the first communications satellite. It was an experimental project and a joint venture for the United States, the United Kingdom and France. It was capable of relaying television signals, and on July 23, 1962 the first satellite television broadcast showed pictures of the Statue of Liberty, the Eiffel Tower and part of a baseball game. These images were broadcast by all the major US networks, across western Europe by Eurovision, and in Canada.

To make communications satellites more useful, they needed to be put in geosynchronous orbit so they would stay in one place in the sky, in effect orbiting the Earth at a velocity equal to the rotation of the Earth. The idea for geostationary orbit is often credited to Arthur C. Clarke (1917–2008) who published an article on the topic in 1945, but he was not involved in actual satellite development. The first geosynchronous satellite was Syncom 2 (synchronous communication satellite) and the first completely geostationary satellite was Syncom 3. Syncom 3, orbiting above the International Dateline, was notable for providing broadcasts of the 1964 Olympic Games from Tokyo to North America. It also provided radio links for the US military during the Vietnam War.

One of the most ambitious NASA projects was the Space Shuttle, which was a low Earth orbit launch and return vehicle. The first shuttle, *Columbia*, was flown on April 12, 1981. Four operational shuttles were built: *Columbia*, *Challenger*, *Discovery* and *Atlantis*. Together, they flew 135 missions, the majority of which were related to deploying satellites. On April 4, 1983, *Challenger* broke apart shortly after takeoff because of the failure of an O-ring, while *Columbia* was destroyed during re-entry by a rupture in its heat shielding on February 1, 2003. The last shuttle flight was made by *Atlantis* on July 8, 2011.

Satellites have contributed far more to scientific discovery than the missions to the Moon, with observatories in space such as the Hubble Telescope and the Fermi Gamma-ray Space Telescope, while weather and earth science satellites provide vast amounts of data. Pointing the observational eyes of satellites toward the ground, numerous spy satellites observe and listen in on every corner of the Earth. For the average person, communications and geopositioning (Global Positioning System [GPS]) have had the greatest impact, ranging from global telecommunications to Google Earth images to

Table 12.1 Satellites by country. In 2017, there were 1,738 operational satellites.

USA 803	China 204	Russia 142	All other 589

navigation systems built into our automobiles. Satellites are necessary for tracking air traffic and container ships, looking for lost cities and harvesting crops.

⚙ Humans as Technology

The extent of our technological life is on the threshold of a new level of complexity as we face the potential to turn ourselves into artifacts. In 2003, the Human Genome Project released its completed map of the human genome. This "map" was essentially a listing in order of all the base pairs (adenine to thymine, guanine to cytosine) that make up human DNA. Although the structure of DNA had been identified in 1953 by James Watson and Francis Crick, the vast and complex process of identifying the component parts had to wait until biochemical tools and computer systems made the project possible. In a very real sense, the genome map was a process of reverse engineering a biological machine, and the information it provided is making the direct manipulation of life at the cellular level possible. Dozens of genomes, from bacteria to blue whales, are being sequenced. Although the idea of "one gene, one characteristic" has been replaced by a much more complex model of gene activity, the conception of the organism as a machine has never been more apparent. Our technology and our biology are intersecting. At a pragmatic level, we are beginning to see ourselves as machines; not as the collection of levers, pistons and hinges of the automatons of the Renaissance, but as a complex self-powered bio-electromechanical system.

If we are machines, it follows that we can be fixed, upgraded and modified. We can potentially be immortal, or at least very long-lived, free of the deterioration of age and the ravages of disease, and transforming the limitations of our current physical bodies. Even if the genetic revolution does not produce these amazing results next week or next year, just the fact that we have the tools to directly manipulate DNA creates a new world of technology.

The genetics goldrush was spurred by a number of landmark cases about patents. As a general rule, patents have not been granted for living organisms because breeding was seen as a natural process even if it was selective breeding to gain specific characteristics. This changed in 1971 when geneticist Ananda Chakrabarty, working for General Electric, applied for a patent on a bacterium that he had modified to consume and break down crude oil. This was proposed as a way of cleaning up oil spills. The US Patent and Trademark Office (USPTO) turned down the patent and triggered a legal fight that went to the Supreme Court. In the case of *Diamond v. Chakrabarty* (Diamond was the head of the USPTO), the Supreme Court ruled that Chakrabarty could patent a modified bacterium. The Patent Office then rewrote its patent rules to allow for patents on genetic material.

One of the first animals patented was the Harvard mouse, also known as OncoMouse. This was a genetically modified mouse that was designed to aid cancer researchers by being susceptible to cancer. The USPTO issued a patent for the Harvard mouse in 1988, but there was a great deal of controversy about the patenting of a higher organism. The European Patent Office eventually issued a patent in 2006, but Canada refused to patent the mouse.

Researchers at universities and private companies began to patent genes for everything from bacteria to male pattern baldness. A limit was placed on human gene patents when the US Supreme Court ruled in 2013 on the case of *Association for Molecular Pathology v. Myriad Genetics*. The court found that simply isolating a human gene was not sufficient for a patent on that gene, since a patent must be for something new and not already existing in nature. Synthetic human genes, however, could be patented.

The implications of treating genes and even whole animals in the same way we dealt with the telephone or the creation of rubber are still being explored today. Many people are concerned that our ability to create new organisms is moving so fast that our moral and social systems cannot keep up.

Nano-world

The origin of tool use gave us the ability to make better use of the material world. In the future, it is likely that the creation of new materials will be at the heart of new technologies. Advances in metallurgy, ceramics, superconductors and new solid-state electronics are all being investigated today, but perhaps the greatest area of research is in the creation of carbon fiber materials. Carbon is at the heart of nanotechnology. The roots of nanotechnology are in the studies of colloids (materials in the 5 to 200 nanometer range) started in the late nineteenth century, but the modern investigation is often attributed to a talk in 1959 entitled "There's Plenty of Room at the Bottom" given by the physicist Richard Feynman (1918–88). He described what it might be possible to do with materials if they could be manipulated at the atomic or molecular level. In 1985, Robert Curl, Harold Kroto and Richard Smalley discovered the three-dimensional C_{60} molecules, known now as buckminsterfullerene because there was a resemblance to the geodesic domes created by Buckminster Fuller (1895–1983). The spheres could be transformed into tubes, and these offer some tantalizing possibilities for future materials, ranging from new batteries to super-strong materials. By manipulating the carbon chains, other atoms could be added and the properties controlled down to the atomic level. Eric Drexler (b. 1955) promoted the importance of nanotechnology in 1986 with his book *Engines of Creation: The Coming Era of Nanotechnology*. In 2010, the Nobel Prize in Physics was awarded to Andre Geim (b. 1958) and Konstantin Novoselov (b. 1974) for their work on graphene, a two-dimensional form of carbon that has many potential applications, from new types of electronics to controlled permeability membranes. Today, companies and the governments of the industrialized nations are pumping billions of dollars into research on nanotechnology.

One of the most interesting ideas that the new materials might make possible is the space elevator. The idea originated with Konstantin Tsiolkovsky (1857–1935), who in 1895 contemplated making a tower like the Eiffel Tower that would reach 35,790 km to orbit. The advantage of such a structure would be very low-cost movement of materials from the ground to orbit, and a way to essentially slingshot supplies and vehicles into space. The idea was addressed for years, sometimes seriously and sometimes as a kind of humorous thought experiment, and appeared in a number of science fiction novels such as Arthur C. Clarke's *The Fountains of Paradise* (1979) and Kim Stanley Robinson's *Red Mars* (1993). The creation of nanocarbon tubes offered the possibility of materials strong enough to actually build a space elevator and serious research is being done on it by NASA in the United States and by the Japanese government. Such an engineering achievement could change human culture as travel to space becomes economical on a very large scale.

Printing the Future

The digital age has also made possible the creation of three-dimensional printing. This technology offers the potential to reduce the cost of material objects enormously. The beginning of 3-D printing started when Hideo Kodama (b. 1950) at the Nagoya Municipal Industrial Research Institute in Japan created a way to build up objects using a liquid plastic that hardened on exposure to ultraviolet light. This was followed in 1984 by the invention of stereolithography by Charles Hull (b. 1939). 3-D printing systems have been created for plastics, ceramics, metals, food, carbon fiber and plant fiber. With 3-D printing, there would be no need for most of the industrial infrastructure created by the Industrial Revolution and less need for delivery networks; physical location would mean almost nothing. Objects could be both mass-produced and completely custom-made at the same time. Shoes that are a perfect fit, a bicycle that matches your physical dimensions or chocolates in the shape of famous sports cars could all be produced in your home or at a nearby print shop. One potential use would be in space colonization. Instead of carrying all the thousands of tools, component parts and machines that would be necessary to establish a base on Mars, many of the objects could be printed when needed.

Digital technology and 3-D printing could transform material culture. The possibilities are only limited by the imagination and physical restraints of the base materials. This would look like magic to people only a couple of generations ago.

The Internet of Things: Harbinger of the Golden Age of Humanity?

In 2014, Jeremy Rifkin in *The Zero Marginal Cost Society* argued that the "internet of things" (IoT) was reducing the marginal cost of production and leading us to a future of great intellectual and material wealth at very low cost. Marginal cost is the cost of producing one additional unit of an object or service. In other words, it may cost millions of dollars to design and create a new automobile, but mass producing all subsequent units is far less expensive and tends to get even cheaper the more copies are made. Products such as music, movies, books and computer apps have already been affected by the shift to a digital economy based on the almost zero marginal cost of delivering the product to the end user. For example digital music, either downloaded or streamed, is nothing more than a string of numbers sent to some electronic device through the internet. Rifkin's argument goes beyond the low cost of information-based products to include physical objects. By combining information technology with automation in production, the marginal cost of physical objects is reduced almost to the cost of the materials needed to build the object. Overhead costs, management and transportation are also reduced by digital systems that manage orders, inventory, logistics, sales and delivery. It is possible today to order things such as disposable razors that are untouched by human hands until the parcel delivery person puts them in your mailbox.

This idea about the production of goods and services gets combined with the IoT because we are connecting everything to the internet. Traffic lights, buses, running shoes, refrigerators, door locks, clothing, home lighting and footballs have all been linked to the internet. It costs so little to track what we do, what we watch, our interests, where we travel and what we buy that goods and services can be individually created almost instantly. Add 3-D printing to the production system and the degree of customization jumps even further. Imagine perfectly fitting shoes and clothes, a refrigerator that orders food for you, cars that drive themselves, street signs that give you directions and clothing that can tell if you are in danger of having a heart attack.

While this science fiction world appeals to many people, there will be social consequences. In the short term, it has already become easier to track people and monitor their behavior, shopping habits and private communication. Cyber bullying and identity theft are serious problems in the connected world. Fake news, cyber manipulation of popular ideology, and fear of governments and corporations invading our privacy are not simply some kind of Luddite paranoia. There have already been shifts in employment, particularly the disappearance of retail jobs and the decline in factory work. There is also the potential to create a widening technology gulf between the connected and the unconnected. Promoters of the IoT say it will be like the printing press, which put some scribes out of business but lifted up millions of people by spreading literacy. Detractors wonder if we will trade freedom for material bounty in an Orwellian world of constant surveillance.

Artificial Intelligence: Technology beyond Human Control?

In the longer term, to manage the digital future we will increasingly look to machine intelligence to run the complex systems, and systems within systems, that are developing today and will become more complex in the era of the IoT. We can also be concerned about the effect of technology on our communities, the environment and the disparity between rich and poor, and, given the growth in genetic technology, we can even be concerned that we are in danger of replacing ourselves with Humans 2.0. We both depend on our technology and are nonetheless uneasy about it. There has been a growing concern that the artificial intelligence we need to run our systems could in effect reach a point of self-replication and design; that digital devices could not only become too complex for us to control or understand, but that such machines might displace us in the way that humans have displaced or driven to extinction many plants and animals. This moment of change has been labeled "the Singularity" (see below). The physicist Stephen Hawking (1942–2018) said:

The development of full artificial intelligence could spell the end of the human race. Once humans develop artificial intelligence, it will take off on its own and redesign itself at an ever-increasing rate. Humans, who are limited by slow biological evolution, couldn't compete and would be superseded. (Cellan-Jones 2014)

This is not a new concern. Fear about the changes caused by technology goes back at least to the time of Plato, and one of the most famous explorations of the role of science and technology was Mary Shelley's *Frankenstein or the Modern Prometheus* (1818). After Charles Darwin published *On the Origin of Species by Means of Natural Selection* in 1859, the novelist Samuel Butler explored the idea of evolution in technology leading to machines that achieve "mechanical consciousness" by means of Darwinian evolution. He also followed this idea in his book *Erewhon* (1872), presenting the idea that, compared to biological evolution, the evolution of machines from ancient times to his time had been incredibly quick. Long before the computer age, Butler was wondering if machines could begin to sense their environment, respond to it, and eventually become self-aware.

The economist W. Brian Arthur has suggested that technology *is* an evolving system, although not following the Darwinian model (Arthur 2009). Technology, according to Arthur, is "autopoietic," meaning self-creating, although to this point it has not gained agency – it still requires an external agent (humans) to direct or enable new developments. Arthur points out that the appearance of new technology primarily builds upon existing technology and follows a rational path based on optimization of subsystems, components and networks. One of the best examples is the mobile phone, which was an intersection of radio technology with telephony and computers. Another technology, the Global Positioning System (GPS) required electronic clocks, radio receivers and transmitters, satellites and computers to exist. Each of these technologies existed before (and was itself a combination of prior technologies), but their combination resulted in a new kind of technology, not an incremental change to any of the existing devices. The utility of GPS made it a natural addition to mobile phone technology, further combining systems to produce a new device.

Arthur's idea also helps to explain the problem of incrementalism in technology and the appearance of new technologies. For example, no amount of incremental improvement in the technology of piston engines for propeller-driven aircraft can push the capability of prop planes beyond certain physical limitations. It is also the case that incremental improvements in propeller-drive aircraft cannot lead to the creation of a jet airplane. Only by combining the ideas for a turbine and a rocket can you conceive of a jet engine.

Taken to its extreme, Arthur's analysis makes people just one of many factors in the environment of the autopoietic technologies. We can expect, in Arthur's model, that *humans* will be subjected to combinatorial technological change through genetics, bio-electronics and electronic additions in exactly the same process as has happened to other forms of technology.

The philosophy of increasingly autonomous technology contains within itself a kind of arrow of direction, from simple devices to complex, with no end point to the complexity except the physical boundaries of nature. Since nature allows the complex "machine" of the human mind and consciousness, there would be no reason not to expect that Butler's "machine consciousness" can and, unless something interrupts the autopoeitic process, *will be* created. This has been labeled "the Singularity," to denote the point at which intelligent computer systems become autonomous and trigger runaway technological development not mediated by people. The term may have been used first by John von Neumann, one of the fathers of modern computing, but it was popularized by Vernor Vinge (b. 1944), professor of computer science and science fiction author, in his 1993 essay "The Coming Technological Singularity." Ray Kurzweil (b. 1948), inventor and computer scientist, is one of the strongest proponents of the Singularity coming in the near future, basing his belief in part on Moore's law about the exponential rise in computing power and the idea

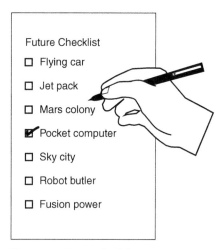

Figure 12.1 Future checklist.

that mapping brains will allow us to emulate self-awareness, creativity and other human characteristics.

The belief in the coming Singularity is a very strong version of technological determinism and requires us to accept the premise that processing power will eventual equal agency, or the ability for a machine to become self-aware and self-directing. Critics point out that there are a number of flaws in the Singularity argument, including the physical limits of chip construction and limits to resources (particularly energy). There is no actual mathematical model of "consciousness" and therefore no way to create a program (which is by its very nature a mathematical process) of consciousness. Moore's law is not a physical law, and the idea that mapping a brain would allow us to emulate a brain is unproven at best. There will be incredible change in the future if even a portion of the technology research being carried out today can be turned into usable products and services, but futurists have been predicting the imminent arrival of AI for more than fifty years. On a more philosophical level, we may fear a future world resembling the dystopias of the movies *Terminator* and *The Matrix*, but we could equally suppose that our future technology would do what most of our technology has done in the past – make life easier for us by solving real-world problems (Figure 12.1)

Considerations of Benefit and Loss

The debate over the possibility of the technological Singularity leads to a deep philosophical question, one that comes from our long love–hate relationship with technology: Has technology been good or bad for humanity? When new technology is introduced, there are always winners and losers, and sometimes who the winners and losers are is not immediately obvious. In the case of the Luddites, it was obvious that the mill owners, as a group, benefited from the new industrial system, and the weavers lost. In the longer term, the poor of Britain were far better off than their ancestors, and certainly lived lives that were far superior to the poor of most of the rest of the world. People in Britain and the other early industrial countries lived longer, were healthier, were more secure in person and property, were better educated and had more choices. The aristocracy, while still a powerful group to this day, nonetheless experienced a long, slow decline in political power and status and, unless they had allied themselves with the industrial and managerial class through marriage or careful investment, a decline in monetary position as well.

If you are Jacques Ellul, the material benefits from technology beyond a certain point do not outweigh the damage to our humanity and perhaps to our souls. If you are Neil Postman, you see technology as having three historical phases: tool use until about the time of the Industrial Revolution; technology until about the time of the invention of the computer; and a new era of technopoly, when systems are self-generating and so complex that only experts can even understand them. While the first two epochs are manageable by individuals and society, the last – technopoly – represents a loss of human autonomy to the systems we have created, and should be avoided. For Kirkpatrick Sale, the defeat of the Luddites was a symbol of the death of traditional, human-centered society and no amount of computing power can restore the communities that grew up when technology was directly under the control of the people.

In defense of the modern world, Sir Kenneth Clark, best known for his sweeping documentary television series *Civilization* (1969), argued that almost all of those things we think of as human virtues – compassion, human rights, a sense of humanity that extended beyond our family, tribe or nation-state – became common ideas only in the modern era and were global objectives only after the Industrial Revolution. Civilization, for Clark and many others, is the embodiment of the belief that tomorrow can be better than today. In the earliest civilizations, that meant the reduction of want: providing food, shelter, security and spiritual support. But as our intelligence gave us more control over nature, civilization also offered us freedom and the opportunity to explore our capabilities not just related to basic survival. These include the fine arts and entertainment, philosophy, education beyond basic survival skills, and, more than anything else, the time to enjoy the benefits of civilization, and contemplate and ask the big questions about life, the universe and everything – as Douglas Adams so nicely put it in the *Hitchhiker's Guide to the Galaxy* (1983).

This book has argued that technology is not just the physical objects that have been invented to solve real-world problems, but the systems that allow us to create and use technology. In fact, some kinds of technology such as government or education have no physical object, but are still powerful forms of technology that allow other forms of technology to function. We have created networks of technology and networks of knowledge to support them. Technology has made us the dominant species on the planet, but has also made us increasingly dependent on our technologies to survive. Ultimately, we are our technology. As individuals, we each interact with tools and devices to a greater or lesser extent, but as a society we have created and are dependent on the systems that allow us to live and thrive. Introduce a new technology, and society is transformed, even if the transformation is very subtle. There are good reasons for society to be cautious about adopting new technologies, and we must not be fooled into believing that the current high value placed on invention will continue indefinitely. If history is any guide, there will be periods when innovation is frequent and others when innovation is less welcome. Making good choices for the present and the future depends on our ability to put into perspective the benefits and problems posed by technology. Knowing that we have faced these issues many times in the past will help us make those decisions.

1 Has technology become indistinguishable from magic?
2 Given that we will soon be able to genetically and digitally enhance ourselves, would you get such enhancements? Would you enhance your children?
3 Would you agree with Neil Postman that we are losing our control of society to technology or with Kenneth Clark that the great characteristic of modern society has been the triumph of human virtues?

✿ FURTHER READING

The problems and benefits of technology are often overshadowed by social anxiety so it is useful to think about history in the clearest and most historical sense. This is the aim of Steven Pinker's *The Better Angels of Our Nature: Why Violence Has Declined* (2012) which, while not specifically about technology, shows that the current state of the world is the way it is because of technology. Peter J. Bowler, better known for his work on the history of biology, has produced a fascinating look at what people expected to have happen in the future: the future is largely based on more and better technology in *A History of the Future: Prophets of Progress from H.G. Wells to Isaac Asimov* (2017). A view of the situation from someone skeptical of the benefits of technology comes from Kirkpatrick Sale, in his *After Eden: The Evolution of Human Domination* (2006). The most important problem related to technology today is climate change. Many sources are available, but the Union of Concerned Scientists is a good place to start, at www.ucsusa.org, while the Intergovernmental Panel on Climate Change at www.ipcc.ch offers a more advanced portal to climate change data.

Bibliography

A History of Education New York: Films Media Group, 2006.

Anthony, David W. *The Horse, the Wheel, and Language: How Bronze-Age Riders from the Eurasian Steppes Shaped the Modern World*. Princeton: Princeton University Press, 2007.

Arthur, W. Brian. *The Nature of Technology: What It Is and How It Evolves*. New York: Free Press, 2009.

Astill, Grenville G. and John Langdon. *Medieval Farming and Technology: The Impact of Agricultural Change in Northwest Europe*. Leiden: Brill, 1997.

Babbage, Charles. *Economy of Manufactures and Machinery*, 4th edn. London: Charles Knight, 1835.

Banerjee, Manabendu and Bijoya Goswami (eds.). *Science and Technology in Ancient India*. Calcutta: Sanskrit Pustak Bhandar, 1994.

Bar-Yosef, Ofer. "The Upper Paleolithic Revolution," *Annual Review of Anthropology* 31 (2002), 363–93.

Beauchamp, Christopher. "Who Invented the Telephone? Lawyers, Patents, and the Judgments of History," *Technology and Culture* 51.4 (2010), 854–78.

Bessemer, Henry. *Henry Bessemer F.R.S.: An Autobiography*. London: Offices of Engineering, 1905.

Blackman, Deane R. "The Volume of Water Delivered by the Four Great Aqueducts of Rome," *Papers of the British School of Rome* 46 (1978), 52–72.

Boot, Max. *War Made New*. New York: Gotham Books, 2006.

Bowler, Peter J. *A History of the Future: Prophets of Progress from H. G. Wells to Isaac Asimov*. Cambridge: Cambridge University Press, 2017.

Brewer, Priscilla J. *From Fireplace to Cookstove: Technology and the Domestic Ideal in America*. Syracuse, NY: Syracuse University Press, 2000.

Brockliss, Laurence and Nicola Sheldon. *Mass Education and the Limits of State Building, c. 1870–1930*. New York: Palgrave Macmillan, 2012.

Brunt, Liam. "New Technology and Labour Productivity in English and French Agriculture, 1700–1850," *European Review of Economic History* 6.2 (2002), 263–7.

Bruun, Mette Birkedal. *The Cambridge Companion to the Cistercian Order*. Cambridge: Cambridge University Press, 2012.

Burns, R. W. *Communications: An International History of the Formative Years*. IET History of Technology 32. London: Institution of Engineering and Technology, 2004.

Cagliarini, Adam and Anthony Rush. "Economic Development and Agriculture in India," Reserve Bank of Australia, *Bulletin* (June 2011). www.rba.gov.au/publications/bulletin/2011/jun/3.html.

Campbell-Kelly, Martin and William Aspray. *Computer: A History of the Information Machine*. Boulder, CO: Westview Press, 2014.

Cowan, Ruth Schwartz. "How We Get Our Daily Bread, or the History of Domestic Technology Revealed," *OAH Magazine of History* 12.2 (1998), 9–12.
More Work for Mother: The Ironies of Household Technology from the Open Hearth to the Microwave. New York: Basic Books, 1983.

Cellan-Jones, Rory. "Stephen Hawking Warns Artificial Intelligence Could End Mankind." *BBC News* December 2, 2014. www.bbc.com/news/technology-30290540.

Chandler, Tertius, Gerald Fox and H. H. Winsborough. *3000 Years of Urban Growth*. New York: Academic Press, 1974.

"Charles Goodyear," *Scientific American Supplement* 787, January 31, 1891.

Clarke, Arthur C. *Profiles of the Future*. SF Gateway, 1973.

Conard, Nicholas J., Maria Mainat and Susanne C. Münzel. "New Flutes Document the Earliest Musical Tradition in Southwestern Germany," *Nature* 24 (June 2009),1–4.

Crowell, Benedict and Robert Forrest Wilson. *The Armies of Industry*. New Haven, CT: Yale University Press, 1921.

Dash, Sean. *The Manhattan Project*. Digital documentary. New York: A & E Television, 2002.

Doyle, Peter. *World War I in 100 Objects*. New York: Plume, 2014.

Drexler, Eric. *Engines of Creation: The Coming Era of Nanotechnology*. New York: Anchor Books, 1986.

Easterbrook, Gregg. "Forgotten Benefactor of Humanity," *Atlantic Monthly*, January 1, 1997.

Ebrey, Patricia Buckley. *The Cambridge Illustrated History of China*. Cambridge: Cambridge University Press, 1996.

Edgerton, David. *The Shock of the Old: Technology and Global History since 1900*. Oxford: Oxford University Press, 2007.

Eisenstein, Elizabeth L. *The Printing Press as an Agent of Change: Communications and Cultural Transformations in Early-Modern Europe*. Cambridge: Cambridge University Press, 2009.

Ellul, Jacques. *The Technological Society*. New York: Vintage Books, 1964.

Farey, John. *Treatise on the Steam Engine: Historical, Practical, and Descriptive*. London: Longman, Rees, Orme, Brown and Green, 1827.

Ferrer, Margaret Lundrigan and Tova Navarra. *Levittown: The First 50 Years. Dover*, NH: Arcadia, 1997.

Films for the Humanities & Sciences. *The Luddites*. Digital video. New York: Films Media Group, 2007.

Frader, Laura Levine. *The Industrial Revolution: A History in Documents*. Oxford: Oxford University Press, 2006.

Franklin, Ursula M. *The Real World of Technology*. Toronto: Anansi, 1999.

Franssen, Maarten, Gert-Jan Lokhorst and Ibo van de Poel. "Philosophy of Technology," in *Stanford Encyclopedia of Philosophy* (plato.stanford.edu).

Gies, Frances and Joseph Gies. *Cathedral, Forge, and Waterwheel: Technology and Invention in the Middle Ages*. New York: HarperPerennial, 1995.

Groves, L. R. *Now It Can Be Told: The Story of the Manhattan Project*. New York: Harper and Row, 1962.

Hacker, Barton C. (ed.). *Astride Two Worlds: Technology and the American Civil War*. Washington, DC: Smithsonian Institution Scholarly Press, 2016.

Hardin, Garrett. "The Tragedy of the Commons," *Ekistics* 27.160 (1969), 168–70.

Harris, Richard and Peter Larkham. *Changing Suburbs: Foundation, Form and Function*. London: Routledge, 1999.

Hartwell, R. M. "Was There an Industrial Revolution?" *Social Science History* 14.4 (1990), 567–76.

Hasan, Ahmad Yusuf. *Islamic Technology: An Illustrated History*. Cambridge: Cambridge University Press, 1986.

Hawthorne, Nathaniel. "Fire Worship," in *Hawthorne: Selected Tales and Sketches*, ed. Hyatt Wagganer. New York: Holt, Rinehart & Winston, 1970, 493–501.

Hayden, Brian. "Models of Domestication," in Anne Birgitte Gebauer and T. Douglas Price (eds.), *Transitions to Agriculture in Prehistory*. Madison, WI: Prehistory Press, 1992.

Headrick, Daniel R. *The Tools of Empire: Technology and European Imperialism in the Nineteenth Century*. Oxford: Oxford University Press, 1981.

Heilbroner, Robert L. "Do Machines Make History?" *Technology and Culture* 8.3 (1967), 335–45.

Hodder, Ian. "Women and Men at Çatalhöyük," *Scientific American* 290.1 (2004), 76–83.

Hughes, Thomas P. *Networks of Power: Electrification in Western Society, 1880–1930*. Baltimore, MD: Johns Hopkins University Press, 1993.

Humphrey, John W., John P. Oleson and Andrew N. Sherwood. *Greek and Roman Technology: A Sourcebook*. London: Routledge, 1998.

Ihde, Don. "Has the Philosophy of Technology Arrived? A State-of-the-Art Review," *Philosophy of Science* 71.1 (2004), 117–31.

Innis, Harold. *Empire and Communication*. Victoria, BC: Press Porcepic, 1986.

Isaacson, Walter. *Steve Jobs*. New York: Simon and Schuster, 2011.

Jaques, R. Kevin. *Authority, Conflict, and the Transmission of Diversity in Medieval Islamic Law*. Leiden: Brill, 2006.

Jewitt, Llewellynn. *The Wedgwoods: Being a Life of Josiah Wedgwood*. London: Virtue Brothers, 1865.

Kennedy, John F. *Special Message to Congress on Urgent National Needs*. May 25, 1961.

Landels, John G. *Engineering in the Ancient World*. Berkeley, CA: University of California Press, 2000.

Le Corbusier. *Vers une architecture*. Paris: G. Crès et Cie, 1923.

Ledeburg, Adolf. *Manuel de la métallurgie du fer*, vol. 1. Paris: Librairie Polytechnique Baudry et Cie, 1895.

Licklider, J. C. R. "Man–Computer Symbiosis," *Transactions on Human Factors in Electronics*, HFE-1 (1960), 4–11.

Lloyd, W. F. *Two Lectures on the Checks to Population*. Oxford: Oxford University Press, 1833.

Martin, Colin and Geoffrey Parker. *The Spanish Armada*. Manchester: Manchester University Press, 1999.

Mather, Ralph. *An Impartial Representation of the Case of the Poor Cotton Spinners in Lancashire &c: with a Mode Proposed to the Legislature for Their Relief*. 1780.

Maxim, Hiram. *Letter to the editor, "How I Invented Maxim Gun,"* *New York Times*, November 1, 1914.

McKenna, Phil "Fossil Fuels Are Far Deadlier than Nuclear Power," *New Scientist*, March 26, 2011.

McNeill, William Hardy. *The Pursuit of Power: Technology, Armed Force, and Society Since A.D. 1000*. Chicago, IL: University of Chicago Press, 1984.

Merwood-Salisbury, Joanna. *Chicago 1890: The Skyscraper and the Modern City*. Chicago, IL: University of Chicago Press, 2009.

Moore, Gordon. "Cramming More Components onto Integrated Circuits," *Electronics* 38.8 (1965), 114–17.

Moore, John. *Jane's Fighting Ships of World War I*. London: Studio Editions, 2001. www .naval-history.net/WW1NavalDreadnoughts.htm.

Mumford, Lewis. *Technics and Civilization, with a New Introduction*. New York: Harcourt, Brace & World, 1963.

Murdock, David and Bonnie Brennan. *Iceman Reborn*. Digital documentary. PBS Nova, 2016.

Needham, Joseph (ed.). *Science and Civilisation in China*, vols. 1–7. Cambridge: Cambridge University Press, 1954–86.

The New Cambridge History of India. Cambridge: Cambridge University Press, 1987–2005.

Nocks, Lisa. *The Robot: The Life Story of a Technology*. Westport, CT: Greenwood Press, 2007.

PBS American Experience. *Henry Ford*. Digital collection of images, film and articles.

Pinker, Steven. *The Better Angels of Our Nature: Why Violence Has Declined*. New York: Penguin Books, 2012.

Postman, Neil. *Technopoly: The Surrender of Culture to Technology*. New York: Vintage Books, 1993.

Rawley, James A. and Stephen D. Behrendt. *The Transatlantic Slave Trade*. Lincoln: University of Nebraska Press, 2005.

Reed, Cameron. "From Treasury Vault to the Manhattan Project," *American Scientist* 9.1 (2011), 40–7.

Rees, Abraham. *The Cyclopædia*. 1820.

Reynolds, Terry S. "Medieval Roots of the Industrial Revolution," *Scientific American* 251.1 (1984), 122–31.

Riello, Giorgio. *The Spinning World: A Global History of Cotton Textiles, 1200–1850*. Oxford: Oxford University Press, 2009.

Romer, John. *The Great Pyramid: Ancient Egypt Revisited*. Cambridge: Cambridge University Press, 2007.

Ronalds, Francis. *Descriptions of an Electrical Telegraph and of Some Other Electrical Apparatus*. London: R. Hunter, 1823.

Rybczynski, Witold. *One Good Turn: A Natural History of the Screwdriver and the Screw*. New York: Scribner, 2000.

Sale, Kirkpatrick. *After Eden: The Evolution of Human Domination*. Durham, NC: Duke University Press, 2006.

Sauer, Carl O. *Agricultural Origins and Dispersals*. New York: American Geographical Society, 1952.

Schafer, Dagmar. *The Crafting of the 10,000 Things*. Chicago: University of Chicago Press, 2011.

Schieber, Philip. "The Wit and Wisdom of Grace Hopper," *OCLC Newsletter* 167 (March/April 1987).

Schmidt, Klaus. "Zuerst kam der Tempel, dann die Stadt," Vorläufiger Bericht zu den Grabungen am Göbekli Tepe und am Gürcütepe 1995–1999. *Istanbuler Mitteilungen* 50 (2000), 5–41.

Sheffield, Gary. *The First World War in 100 Objects*. London: André Deutsch, 2013.

Shubik, Martin. "The Dollar Auction Game: A Paradox in Noncooperative Behavior and Escalation," *Journal of Conflict Resolution* 15.1 (1971), 109–11.

Steinberg, S. H. and John Trevitt. *500 Years of Printing*. London: British Library, 1996.

Svizzero, Serge and Clement Tisdell. "Theories about the Commencement of Agriculture in Prehistoric Societies: A Critical Evaluation," *Economic Theory, Applications and Issues Working Papers*, August 2014, 1–28.

Understanding Greek and Roman Technology, by The Great Courses, 2013.

White, K. D. *Greek and Roman Technology*. Ithaca, NY: Cornell University Press, 1984.

White, Lynn. *Medieval Technology and Social Change*. Oxford: Oxford University Press, 1962.

Wrangham, Richard. *Catching Fire: How Cooking Made Us Human*. London: Profile Books, 2009.

Zimmer, Carl. *Smithsonian Intimate Guide to Human Origins*. Washington, DC: Smithsonian Books, 2005.

Index